WITHDRAWN

VNR COMPETITIVE MANUFACTURING SERIES

Product and Process Design
PRACTICAL EXPERIMENT DESIGN by William J. Diamond
VALUE ANALYSIS IN DESIGN by Theodore C. Fowler
A PRIMER ON THE TAGUCHI METHOD by Ranjit Roy
MANAGING NEW-PRODUCT DEVELOPMENT by Geoff Vincent
ART AND SCIENCE OF INVENTING by Gilbert Kivenson
RELIABILITY ENGINEERING IN SYSTEMS DESIGN AND OPERATION by Balbir S. Dhillon
RELIABILITY AND MAINTAINABILITY MANAGEMENT by Balbir S. Dhillon and Hans Reiche
APPLIED RELIABILITY by Paul A. Tobias and David C. Trindad

Manufacturing (hard)
INDUSTRIAL ROBOT HANDBOOK: CASE HISTORIES OF EFFECTIVE ROBOT USE IN 70 INDUSTRIES by Richard K. Miller
ROBOTIC TECHNOLOGY: PRINCIPLES AND PRACTICE by Werner G. Holzbock
MACHINE VISION by Nello Zuech and Richard K. Miller
DESIGN OF AUTOMATIC MACHINERY by Kendrick W. Lentz, Jr.
TRANSDUCERS FOR AUTOMATION by Michael Hordeski
MICROPROCESSORS IN INDUSTRY by Michael Hordeski
DISTRIBUTED CONTROL SYSTEMS by Michael P. Lukas
BULK MATERIALS HANDLING HANDBOOK by Jacob Fruchtbaum
MICROCOMPUTER SOFTWARE FOR MECHANICAL ENGINEERS by Howard Falk

Manufacturing (soft)
WORKING TOWARDS JUST-IN-TIME by Anthony Dear
GROUP TECHNOLOGY: FOUNDATION FOR COMPETITIVE MANUFACTURING by Charles S. Snead
FROM IDEA TO PROFIT: MANAGING ADVANCED MANUFACTURING TECHNOLOGY by Jule A. Miller
COMPETITIVE MANUFACTURING by Stanley Miller
STRATEGIC PLANNING FOR THE INDUSTRIAL ENGINEERING FUNCTION by Jack Byrd and L. Ted Moore
SUCCESSFUL COST REDUCTION PROGRAMS FOR ENGINEERS AND MANAGERS by E. A. Criner
MATERIAL REQUIREMENTS OF MANUFACTURING by Donald P. Smolik
PRODUCTS LIABILITY by Warren Freedman
LABORATORY MANAGEMENT: PRINCIPLES & PRACTICE by Homer Black, Ronald Hart, Orrin Peterson

Materials Management
TOTAL MATERIALS MANAGEMENT: THE FRONTIER FOR COST-CUTTING IN THE 1990S by Eugene L. Magad and John Amos
MATERIALS HANDLING: PRINCIPLES AND PRACTICE by Theodore H. Allegri, Sr.
PRACTICAL STOCK AND INVENTORY TECHNIQUES THAT CUT COSTS AND IMPROVE PROFITS by C. Louis Hohenstein

Contents

Chapter 1 Machine Vision — A Data Acquisition System 1
Chapter 2 Machine Vision Hardware......................9
Chapter 3 Machine Vision Software 23
Chapter 4 Checklist For Machine Vision Applications........47
Chapter 5 Industrial Applications for Inspection............61
Chapter 6 Vision For Industrial Robots 89
Chapter 7 A Summary of Research at SRI International.....107
Chapter 8 A Summary of Research
 at the University of Rhode Island 119
Chapter 9 A Summary of Research at Stanford University ...131
Chapter 10 A Summary of Research at Purdue University.....145
Chapter 11 A Summary of Research
 at the National Bureau of Standards 159
Appendix A Manufacturers of Machine Vision Systems........173
Appendix B Glossary...................................183
Appendix C Bibliography................................195
Index..209

Preface

These are exciting times for one to be involved in manufacturing. Advances in microelectronic technology have brought forth a wealth of new machines and controls to the manufacturing environment. The competitive advantage in business is clearly shifting toward those companies that utilize these tools.

One of the most advanced emerging manufacturing tools is machine vision. U.S. industry has only scratched the surface in applying this powerful tool. While the number of machine vision sytems in use in the United States today is approaching 10,000, it has been suggested that the number of potential applications is in the hundreds of thousands.

Today machine vision is now applied primarily by companies considered to be technology leaders. The next decade will see engineers and managers in virtually every manufacturing company in the United States investigate and apply this technology. This book is a guide for that investigation process. It provides the manufacturing engineer and manager with an overview of the technology, ranging from how machine vision works to how systems are applied.

The book also provides historical background on machine vision. Chapters 7 thru 11 review developments at leading laboratories during the formative years of machine vision technology. Having prepared the engineer or manager with the background to begin to develop specific machine vision applications, the book concludes with a list of addresses of machine vision vendors, who may be contacted to obtain information on current commercial systems.

1
Machine Vision—A Data Acquisition System

Machine vision is not an end unto itself! It represents a piece of the manufacturing/quality control universe. That universe is driven by data related to the manufacturing process. That data is of paramount importance to upper management as it relates directly to bottom-line results. For competitiveness factors top management can not delegate responsibility for quality control. Quality assurance must be built in—a totally integrated function—integrated into the whole of the design and manufacturing process. The computer is the means to realize this integration.

Sophisticated manufacturing systems require automated inspection and test methods to guarantee quality. Methods are available today, such as machine vision, that can be applied in all manufacturing processes: in-coming receiving, forming, assembly, and warehousing and shipping. However, hardware alone should not be the main consideration. The data from such machine vision systems is the foundation for computer integrated manufacturing. It ties all of the resources of a company together—people, equipment and facilities.

It is the manufacturing data that impacts quality, not quality data that impacts manufacturing. The vast amount of manufacturing data requires examination of quality control beyond the traditional aspects of piece part inspection, into areas such as design, process planning, and production processes.

The quality of the manufacturing data is important. For it to have an impact on manufacturing it must be timely as well as accurate. Machine vision systems when properly implemented automate the data capture and can in a timely manner be instrumental in process control. By recording this data automatically from vision systems, laser micrometers, tool probes and machine controllers, input errors are significantly reduced and human interaction minimized.

Where data is treated as the integrator, the interdepartmental data base is fed and used by all departments. Engineering loads drawing records. Purchasing orders and receives material through exercise of the same data base which finance also uses to pay suppliers. Quality approves suppliers and store results of incoming inspection and tests on these files. The materials function stocks and distributes parts and manufacturing schedules and controls the product flow. Test procedures stored drive the computer aided test stations and monitor the production process.

The benefits of such an "holistic" manufacturing/quality assurance data management system include:

Increased productivity:
- reduced direct labor
- reduced indirect labor
- reduced burden rate

increased equipment utilization

increased flexibility

reduced inventory

reduced scrap

reduced lead times

reduced set-up times

optimum balance of production

reduced material handling cost and damage

predictability of quality

reduction of errors due to:
- operator judgement
- operator fatigue
- operator inattentiveness
- operator oversight

increased level of customer satisfaction

Holistic manufacturing/quality assurance data management involves the collection (when and where) and analysis (how) of data that conveys results of the manufacturing process to upper management as part of a factory-wide information system. It merges the business applications of existing data processing with this new function.

It requires a partnership of technologies to maximize the production process to ensure efficient manufacturing of finished goods from an energy, raw material, and economic perspective. It implies a unified systems architecture and information center software and data base built together. This integrated manufacturing, design and business functions computer based system would permit access to data where needed as the manufacturing process moves from raw material to finished product.

Today such a data driven system is possible. By placing terminals, OCR readers, bar code readers and machine vision systems strategically throughout a facility it becomes virtually paperless. For example, at incoming receiving upon receipt of material, receiving personnel can query the purchasing file for open purchase order validation, item identification and quality requirements. Information required by finance on all material receipts is also captured and automatically directed to the accounts payable system.

The material can then flow to the mechnical and/or electrical inspection area where automatic test equipment, vision systems, etc. can perform inspection and automatically record results. Where such equipment is unavailable, inspection results can be entered via a data terminal by the inspector. Such terminals should be user friendly. That is, designed with tailored keys for the specific functions of the data entry operation.

Actual implementation of such a data driven system will look different for different industries and even within the industry different companies will have different requirements because of their business bias. For example, a manufacturer of an assembled product who adds value with each step of the process might collect the following data:

receiving inspection:
    a. total quantity received by part number
    b. quantity on the floor for inspection
    c. quantity forwarded to production stock
    d. calculation of yield

inventory with audit (reconciliation) capability:
    a. ability to adjust, eg. addition of rework

b. FIFO/LIFO
  c. part traceability provisions
  d. special parts

production:
  a. record beginning/end of an operation
  b. ability to handle exceptions—slow moving or lost parts
  c. ability to handle rework
  d. ability to handle expedite provisions
  e. provide work in process by part number, operation
  f. provide process yield data by:
     − part number
     − process
     − machine
  g. current status reporting by:
     − part number
     − shop order number
     − program operation
     − rejection
  h. activity history of shop order in process including rework
  i. shop orders awaiting kitting
  j. shop orders held up because of component shortages
  k. history file for last "X" months
  l. disc and terminal utilization

quality:
  Provide hard copy statistical reporting data (pie charts, bar diagrams, histograms, etc.)

data input devices:
  a. OCR
  b. bar code readers
  c. keyboards
  d. test equipment
  e. machine vision systems

personnel:
  a. quality control inspectors
  b. production operators
  c. test technicians

With appropriate sensor technology the results include unattended machining centers. Machine mounted probes, for example, can be used to set up work, part alignment and a variety of in process gaging operations. Microprocessor-based adaptive control techniques are currently available which can provide data such as:

tool wear

tool wear rate greater or less than desired

work piece hardness different from specification

time spent

percentage of milling vs. drilling time, etc.

Quality assurance can now use CAD/CAM systems for many purposes; for example, to prove numerical control machine programs, and provide inspection points for parts and tools.

After the first part is machined, inspection can be performed on an off-line machine vision system analogous to a coordinate measurement machine using CAD developed data points. This verifies the NC program contains the correct geometry and can make the conforming part. At this point QA buys off the program software. While the program is a fixed entity and inspection of additional parts fabricated for shape conformance is not needed, inspection is required for elements subject to variables: machine controller malfunction, cutter size, wrong cutter, workmanship, improper part loading, omitted sequences and conventional machining operations. This may necessitate sample inspection of certain properties—dimensions, for example, and a 100% inspection for cosmetic properties—tool chatter marks, for example.

The CAD/CAM system can be used to prepare the inspection instructions. Where automatic inspection is not possible, a terminal at the inspection station displays the view the inspector sees along with pertinent details. On the other hand, it may be possible in some instances to download those same details to a machine vision system for automatic conformance verification. CAD systems can also include details about the fixturing requirements at the inspection station. This level of automation eliminates the need for special vellum overlays and optical comparator charts. The machine vision's vellum or chart is internally generated as a referenced image in the computer memory.

While dimensional checks on smaller parts can be performed by fixturing parts on an X-Y table that moves features to be examined under the television camera, larger objects can be similarly inspected by using a robot to move the camera to the features to be inspected or measured. Again, these details can be delivered directly from CAD data.

Analysis programs for quality monitoring can include:

— Histogram which provides a graphic display of data distribution. Algorithms generally included automatically test the data set for distribution, including skewness, kutosis and normality.

— Sequential plots which analyze trends—to tell, for example, when machine adjustments are required.

— Feature analysis to determine how part data compares with tolerance boundaries.

— Elementary statistics programs to help analyze data of workpiece characteristics—mean, standard deviation, etc.

— X-bar and R control chart programs to analyze the data by plotting information about the averages and ranges of sequences of small samples taken from the data source.

A computer aided quality system can eliminate paperwork, eliminate inspection bottlenecks and expedite manufacturing batch flow. The quality function is the driver that merges and integrates manufacturing into the factory of the future.

Bibliography

Koon, Troy and Jung, Dave, "Quality Control Essential for CIM Success," Cadlinc, Inc., Elk Grove Village, Il., Jan-Mar, 1984, pp. 2-6.

Papke, David, "Computer Aided Quality Assurance and CAD/CAM," Proceedings CAM-I Computer Aided Quality Control Conference, May 1982, Baltimore, Md., pp. 23-28.

Bellis, Stephen J., "Computerized Quality Assurance Information Systems," Proceedings CAM-I Computer Aided Quality Control Conference, May, 1982, Baltimore, Md., pp. 107-112.

Bravo, P.F. and Kolozsvary, "A Materials Quality System in a Paperless Factory," Proceedings CAM-I Computer Aided Quality Conference, May, 1982, Baltimore, Md., pp. 113-121.

Gehner, William S., "Computer Aided Inspection and Reporting CAIR," Proceedings CAM-I Computer Aided Quality Conference, May, 1982, Baltimore, Md., pp. 175-182.

Kutcher, Mike and Gorin, Eli, "Moving Data, not paper, enhances productivity," IEEE Spectrum, May, 1983, pp. 84-88.

Kutcher, Mike, "Automating it All," IEEE Spectrum, May, 1983, pp. 40-43.

Barker, Ronald D., "Managing Yields by Yielding Management to Computers," Computer Design, June 1983, pp. 91-96.

Schaeffer, George, "Sensors: the Eyes and Ears of CIM," American Machinist, July, 1983, pp. 109-124.

2
Machine Vision Hardware

Machine vision systems have two primary elements: the camera, which serves as the eye of the system, and a computer video analyser. Lighting is also an important consideration. The camera, which may use a vidicon or solid-state sensor, scans a scene at a rate of 1/60 second. The sensor converts spatial data into a time varying signal where time relates to position and intensity to the analog signal level at that time. This stream of information from the camera, representing a two-dimensional image of the scene, is fed serially into the camera interface, digitizer and processing computer.

The pixel is the individual element in a digitized array, and may refer to the picture elements of either a solid-state camera or vidicon camera.

VIDICONS

Many vision systems analog electrical signals are from conventional closed circuit television cameras with vidicons (also plumbicons and silicon target vidicons). Vidicons have some disadvantages which are leading to their replacement in machine vision systems by solid-state sensors. Vidicons depend on an electron beam scanned across an image target to create a signal electrostatically. That beam is deflected and experiences geometric distortion. Camera tubes also have lag, the trailing-comet-like image produced when a moving light is seen against a dark background. Bright lights can burn and damage a tube target. All of these phenomena can cause distortion of the camera signal, resulting in erroneous information being passed to the video image analyzing computer. In addition, video tube cameras are fragile and subject to damage from shock and vibration.

SOLID STATE CAMERAS

Silicon detectors called charge-coupled devices (CCD's) were invented at Bell Laboratories in 1969 by William Boyle and George Smith. These devices generate an electronic signal proportional to incident light. Silicon is known to absorb photons in the range of 200 to 1100 nm. CCD's are self-scanning, with precision permanently etched in its silicon structure, and transmit signals representing the scene being analyzed in periodic discrete "packets" of information easily understood by the interfaced computer.

Solid-state image sensors are generally enclosed at integrated circuit packages with ground and polished glass or quartz windows. They are usually a monolithic silicon chip that contains the photosensors and associated readout circuitry. Figure 2-1 shows the four basic architectures of a solid-state image sensor. The first structure uses photodiodes as detectors and a digital shift register to sequentially interrogate the photodiodes, and is depicted in the figure as combination A. The second architecture, combination B, is commonly referred to as a charge-coupled device which uses the field-induced photo-detector as the pixel and the analog shift register to shift the information from the pixel to the output terminal. The third structure, shown as combination C, combines the field-induced photo-detector with the digital shift register in an effort to obtain the higher density with existing technology. The final structure uses photodiodes combined with an analog shift register for readout and is commonly called a CCPD or charge-coupled photodiode array.[1]

Operationally, a solid-state image sensor converts incident light to electric charge which is integrated and stored until readout. The integrated charge is directly proportional to the intensity of the light impinging on the sensing elements. Readout is initiated by a periodic start or transfer pulse. The charge information is then sequentially read out at a rate determined by clock pulses applied to the image sensor. The output is a discrete time analog representation of the spatial distribution of light intensity across the array.[1]

Figure 2-2 shows a block diagram of typical image sensor support circuitry and required sensor clocking waveforms. Referring to the video output waveform in Figure 2-2, the darkened pixels 5, 6, 7 and 8 could relate to the diameter of the cable in the sketch. If the analog video is compared to a threshold voltage, then the digital

MACHINE VISION HARDWARE 11

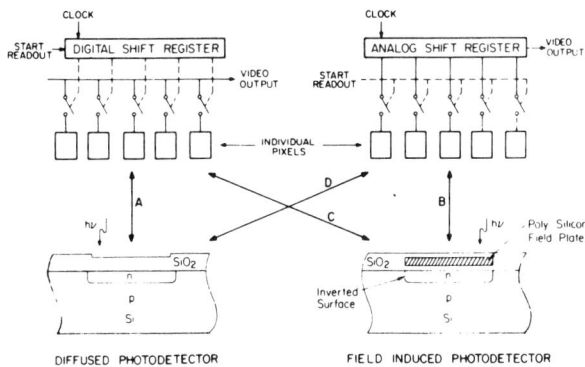

Figure 2-1. Four basic architectures of a solid-state image sensor.[1]

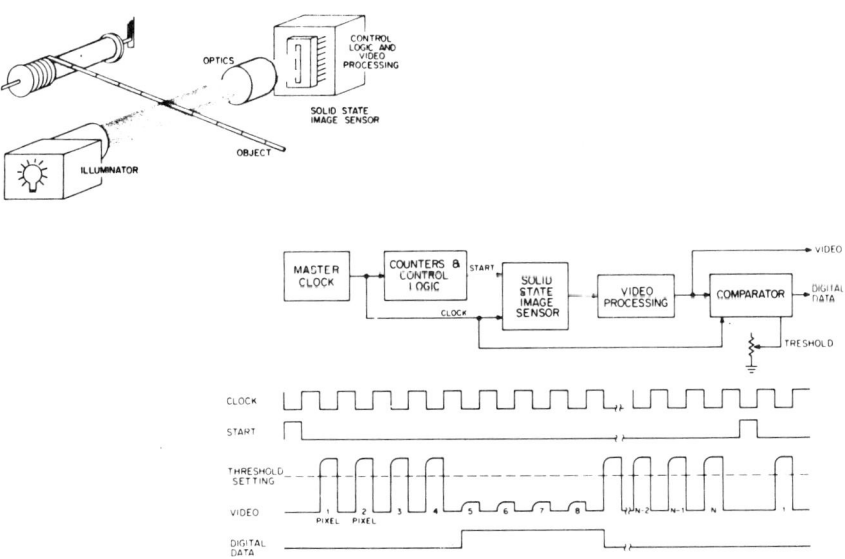

Figure 2-2. Solid-state image sensor drive requirements. The sketch of an object measurement depects a measurement situation which may generate the data shown.[1]

data pulse shown can be generated. By counting the number of clock pulses in the digital data, the number of pixels in the shadow of the cable can be determined. Since the pixels are on precise geometrical centers, the diameter of the cable can be calculated by multiplying the number of elements times their center-to-center spacing times the optical magnification factor.[1]

CCD's are manufactured into matrix, linear and circular arrays. The first two types are of most interest for machine vision systems, producing two-dimensional and one-dimensional images of scenes respectively. A solid state pixel array will generate a representation of an entire scene or a window of a scene. A linear array may be used for objects which are in relative motion to the camera, such as parts moving on a conveyor.

SOLID-STATE MATRIX ARRAYS

Some solid-state matrix arrays are shown in Figure 2-3. Commercially available cameras which utilize these photo-active chips are shown in Figure 2-4.

RESOLUTION

If the digital representation of an image is given by a discrete function $f(i,j)$, $0 = i = M$ and $0 = j = N$, then the resolution of the digital image is given by the values M and N and the number of bits by which the gray scale values of $f(i,j)$ are represented.

Currently, most machine vision systems process a 320 x 480 matrix. This results in a pixel of 1/320 of the field of view, or 0.3%. In applications requiring greater definition, the object must be scanned with a linear array scan camera, in sections with multiple cameras or through a step-and repeat process. A step and repeat scanner can divide the object into sections to be scanned one by one by using X-Y translation. Processing speed, of course, is proportional to both the number of pixels and object positioning time.

MACHINE VISION HARDWARE

Figure 2-3. Examples of solid-state CCD pixel arrays. (*Source: EG&G Reticon and Fairchild*)

Figure 2-4. Commercial cameras which utilize CCDs.

MACHINE VISION HARDWARE

Some typical pixel array sizes used in vision systems are:

Size	Total Pixels
32 x 32	1,024
64 x 64	4,096
100 x 100	10,000
128 x 128	16,384
244 x 248	60,512
256 x 256	65,536
320 x 240	76,800
320 x 480	153,600
320 x 512	163,840

It is noted that most of the pixel array arrangements are binary place values, ($2^6=64$) as found in the preceeding table. The 320 x 480 arrays represent the current upper limits in vision system resolution. An array of 320 x 240 is a typical size for a pixel array because this aspect ratio easily synchronizes with the picture format of existing EIA, RS170 closed-circuit TV cameras.

An example of the visual effect of the number of pixels in a binary image is shown in Figure 2-5.

Looking toward the future of solid-state cameras[5], there is a need for higher resolution (i.e., less than ½ mil between neighboring centers) and precision. There is also a need for improved quality of pixels (i.e., fewer defective elements, higher and more uniform sensitivity, a wider dynamic range of intensity, and antiblooming) and color discrimination (i.e., preferable red, green, and blue). These extended capabilities should be achieved without substantial cost increases as higher density semiconductor manufacturing advances are made.

SCANNING AND PROCESSING TIME

Matrix cameras generally require 1/60 second (16.6 milliseconds) to scan. However, at this speed, only information relating to the presence or absence of an object can be processed. This scanning speed translates to an upper bound of 3,600 objects per second for array camera vision system throughput.

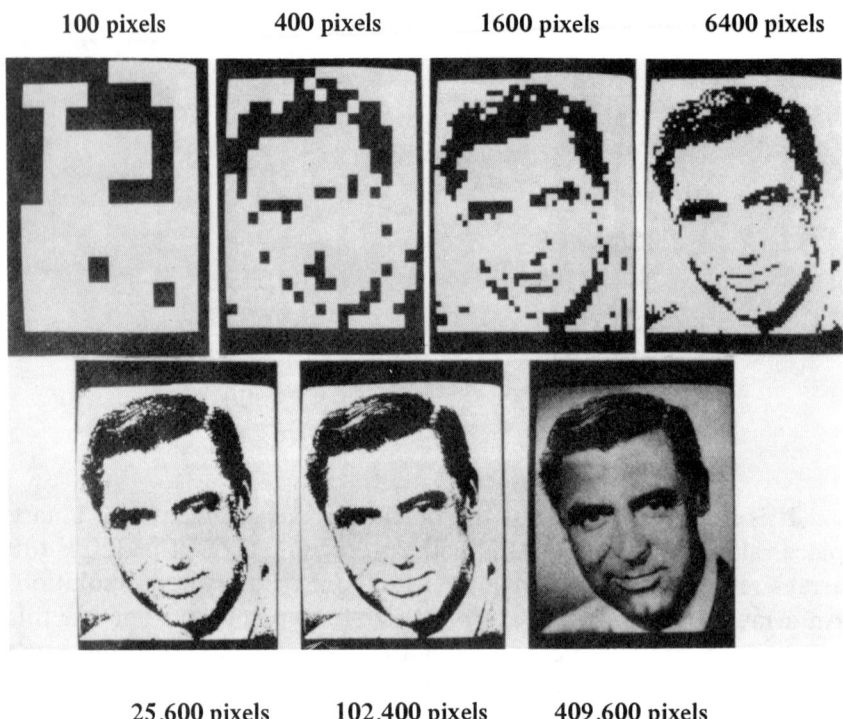

Figure 2-5. Digitized photograph of Cary Grant with various number of pixels.[10]

In most industrial vision analysis applications, more information than simple object presence or absence is required for analysis. To calculate information relating to area, position and orientation, one second would be required for a 65,536 pixel array (256 x 256). The processing time is directly proportional to array size, so ¼ second would be required for a 128 x 128 array, 1/16 second for a 64 x 64 array, and 2.3 seconds for a 320 x 480 array. Thus, slower throughputs are achieved than indicated by the 1/60 second camera scanning rate. Throughputs of 60 to 300 parts per minute are more typical. These speeds are adequate for most industrial applications.

SOLID-STATE LINEAR ARRAYS

Solid-state linear arrays using CCD's are commercially available in sizes ranging from 16 to 4096 pixels. These devices perform repetitive single scans of objects in motion relative to the camera. An example is shown in Figure 2-6, where a connecting rod moves past the viewing station and top-view and side-view linear scans are performed by two linear diode arrays, each scan initiated by a repetitive signal from a position sensor (incremental encoder) that is coupled to the moving conveyor belt.[3] For each scan, values of light intensity at a fixed number of discrete points are measured, converted into electrical signals, and sent to a computer. These signals are either processed in real time or stored in memory until the image of the entire workpiece is obtained for subsequent processing.

With up to 4096 pixels, linear arrays can provide a greater resolution than matrix array cameras, which are generally limited to a maximum of 512 pixels. This higher resolution, however, results in a proportionately higher amount of data to process and associated processing time. Thus, if a 2048 pixel line scan camera and a 256 x 256 pixel array camera both view and analyze a square area, the line scan camera will have a throughput 54 times slower (i.e., approximately 64 seconds) but will provide greater resolution. Resolution can also be enhanced by viewing one edge on an object while referencing the opposite edge to a known coordinate.

LIGHTING

Some of the common illumination sources are tungsten, quartz halogen, quartz iodine, fluorescent, and mercury—or xenon—arc lamps, as well as various flash lamps, lasers and Light Emitting Diode (L.E.D.) sources. The common ways to configure these sources are: front light or spot and back light or spot, as well as collimated back lighting. The line illumination is used with a linear solid-state image sensor while the spot and collimated sources can be used with either the linear or matrix solid-state image sensor. Figure 2-7 illustrates such object lighting techniques as front lighting, rear lighting, spectral illumination, spectral elimination, beam splitting, split mirror, offset shadowing, and collimated light.[1]

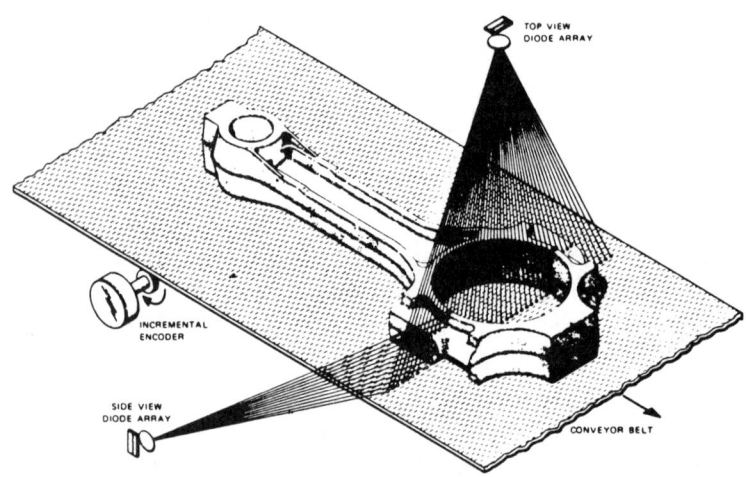

Figure 2-6. Application of linear CCD array.[6]

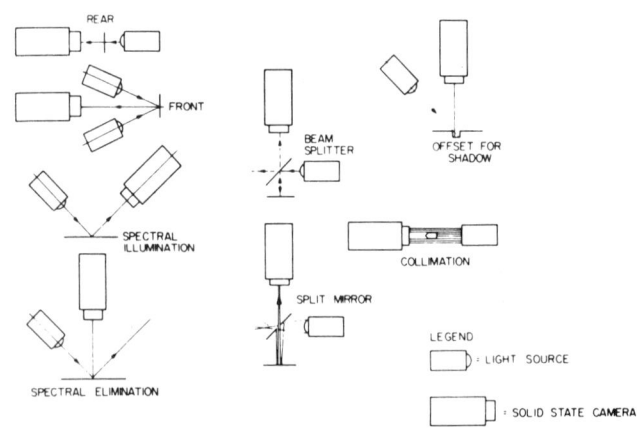

Figure 2-7. Basic lighting techniques.[1]

When possible, rear lighting is preferred since it provides greater image contrast.[7] Front lighting must be used, however, where surface features must be extracted. Light intensity must be sufficient to swamp interferences from ambient sources.

Image contrast is also a constraint.[7] The contrast of the object against its background must be greater than the local lighting variation around a feature of interest. Lighting variations are caused by point light sources and interference from ambient light. Features that the system must extract, such as edges or holes, should be distinguished from the local background by 15 to 25% of the overall image intensity range for reliable detection. Thus, for example, using a brightness scale of 1 to 10, a system can distinguish an edge if local background intensity is at level 3 and the edge is illuminated with at least 4.5. Nominal edge-level intensity should be 5.5 or greater.

Structured light is the use of sheets of light and other projective light configurations to directly determine shape and/or range from the observed configuration that the projected line, circle, grid, etc. makes as it intersects the object.

An example of structured lighting for machine vision is the system developed by General Motors for their CONSIGHT system.[13-16]

A slender tungsten bulb and cylindrical lens are used to project a narrow and intense line of light across the belt surface. The line camera is positioned so as to image the target line across the belt. When an object passes into the beam, the light is intercepted before it reaches the belt surface (Figure 2-8). When viewed from above, the line appears deflected from its target wherever a part is passing on the belt. Therefore, wherever the camera sees brightness, it is viewing the unobstructed belt surface; wherever the camera sees darkness, it is viewing the passing part.

Unfortunately, a shadowing effect causes the object to block the light before it actually reaches the imaged line, thus distorting the part image. The solution is to use two (or more) light sources all directed at the same strip across the belt (Figure 2-8). When the first light source is prematurely interrupted, the second will normally not be. By using multiple light sources and by adjusting the angle of incidence appropriately, the problem is essentially eliminated.

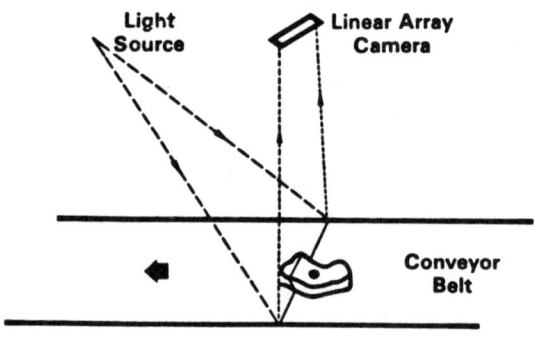

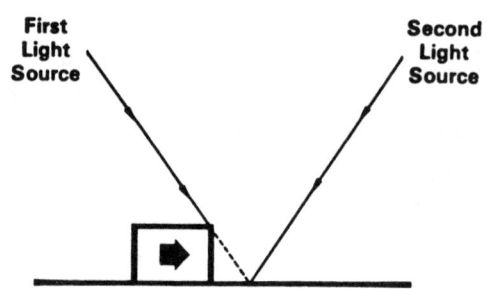

Figure 2-8. CONSIGHT vision system developed by General Motors, showing basic lighting principle and improvement with two light sources.[13-16]

References

[1] Hopwood, Ronald K., "Minicomputers and Microprocessors in Optical Systems, Proceedings of the Society of Photo-Optical Instrumentation engineers, S. 230, 1980, pp. 72-82.

[2] Free, John, "Chips That See," *Popular Science,* January 1982, pp. 61-63.

[3] C.A. Rosen, D. Nitzan, et al., "Exploratory Research in Advanced Automation," Reports 1 through 5, NSF Grants GI-38100X and GI-38100X1, Stanford Research Institute, Menlo Park, California 94025 (December 1973 through January 1976)

[4] G.F. Amelop, "Charge-Coupled Devices," *Scientific American,* February 1974.

[5] Nitzan, David, "Assessment of Robotic Sensors," Proceedings of the First International Conference on Robot Vision and Sensory Controls, IFS Conferences, Ltd (U.K.), 1981.

[6] C.A. Rosen and D. Nitzan, "Some Developments in Programmable Automation," *Proc. IEEE Intercom 75*, (9 April 1975); also *Manufacturing Engineering*, pp. 26-30 (September 1975).

[7] Trombly, John E., "Adding Machine Vision To Assembly Lines," *Machine Design*, November 11, 1982, pp. 78-81.

[8] Trombly, John E., "Recent Applications of Computer Aided Vision In Inspection and Part Sorting," Presented at Robot VI Conference, March 2-4, 1982, Detroit, MI, SME technical paper no. MS82-128.

[9] Grevarter, William B., "An Overview of Computer Vision," National Bureau of Standards, Report No. NBSIR 82-2582, September 1982.

[10] Bradbeer, R., DeBono, P., and Laurie, P., "The Beginners Guide To Computers," Addison-Wesley, 1982.

[11] Nowak, Glenn, "The Advent of Machine Vision Systems," *Manufacturing Engineering*, November 1982, pg. 56-60.

[12] "1982 Robotics Industry Directory," Technical DataBase Corporation, Conroe, Texas, 1982.

[13] Ward, M.R., Rossol, L. and Holland, S.W., "Consight: A Practical Vision-Based Guidance System," Proceedings, 9th International Symposium on Industrial Robots, Washington, DC, March 1979.

[14] Rossol, L. "Vision and Adaptive Robots In General Motors," Presented at the

[15] Holland, S.W., Rossol, L. and Ward, M.R., "Consight-I: A Vision Controlled Robot System for Transferring Parts from Belt Conveyors," *Computer Vision and Sensor-Based Robots*, G.G. Dodd and L. Rossol, Eds. New York: Plenum Press, 1979, pp. 81-100.

[16] Ward, M.R., Rossol, L., Holland, S.W., "CONSIGHT: An Adaptive Robot with Vision," *Robotics Today*, pp. 26-32, Summer, 1979.

3
Machine Vision Software

The recent rapid acceleration in the development of machine vision for industrial applications can be attributed to research in the areas of computer technologies. The first step in vision analysis is the conversion of analog pixel intensity data into digital format for processing. Next, an appropriate computer algorithm is employed to understand the image data and provide appropriate analysis or action. This chapter assesses the various methods for computer processing and analysis of visual information.

HOW COMPUTER RECOGNITION WORKS

An easy-to-understand explanation of the machine vision recognition process was provided by *High Technology*[1], and is reprinted with permission. Reference is made to Figure 3-1. (This description reviews essentially the approach developed by Stanford Research Institute, Stanford University.)

Recognition involves several steps. First, the objects in the television camera image are reduced to their silhouettes by setting all the gray-scale intensities in the object's background to black and all the object intensities to white (1). This silhouetting is done as the image is scanned from the TV camera, using a technique called thresholding. The system assumes that objects of interest will contrast sharply with the background. Thus all object intensities will be above or below a certain level, depending on whether the objects are lit from the front or the back. This threshold value is used to determine which intensities should be set to white and which to black during silhouette formation.

The threshold value itself is determined by the operator during system training. As an aid to threshold determination, the vision

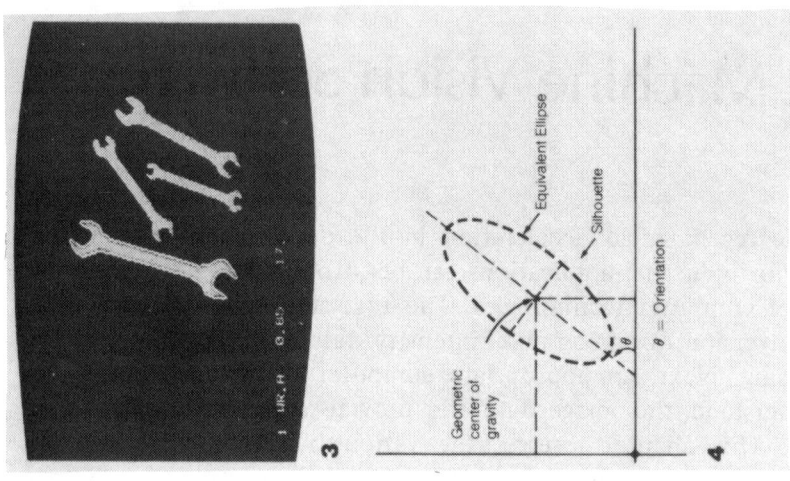
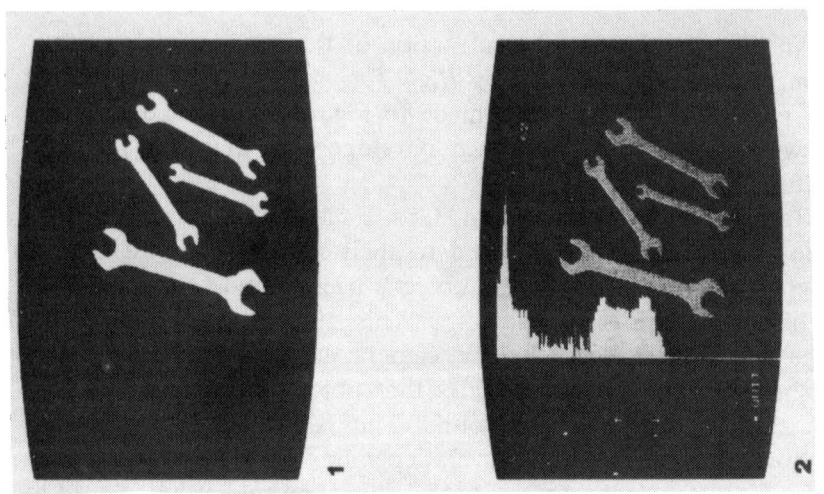

Figure 3-1. Example showing how a vision system sees in two dimensions.[1]

MACHINE VISION SOFTWARE

system displays a histogram of the intensities encountered in a typical scene (2). If the scene has sufficient contrast, the histogram will have two peaks—a dark one for the background and a bright one for the objects. The threshold value will then be the bottom of the "valley" between the peaks.

The outlines of the silhouettes are traces (3). Tracing is done by systematically scanning the image for silhouette edges, starting in the upper left-hand corner and moving line by line to the bottom. (Actually, the system scans a compressed version of the image to minimize scanning time.) The compressed image contains the beginning and end locations of continuous runs of the same grey level—the only information needed to determine silhouette edges. This is called run-length encoding. A transition from black to white (left edge) or white to black (right edge) signals the edges of a white silhouette on a black ground. Whenever the computer encounters an edge point, it determines what edge the point belongs to by examining neighboring points on the line above. It then enters the point's location in a list for that edge. If the point is isolated, however, the system assumes a new silhouette and creates a new list.

Because the computer always scans the image from top to bottom, it can never be sure that the edge it encounters through the scan is not connected further down in the image. For example, if the system first encounters the line of a fork, it will treat the lines as separate objects until it reaches the fork's palm. For this reason, edge lists are always provisional until a scan is completed. Lists are consolidated when edges are found to be connected, which happens when the system discovers a point common to two or more edges.

Next the system computes the location and orientation of the silhouettes. A silhouette's location is defined as its geometric center of gravity; its orientation, as the orientation of an ellipse that has the same area (4) or in terms of some other geometric property.

Finally the system attempts to match the silhouettes to the examples stored in its memory. A close match is considered recognition. The closeness of a match is determined by scoring individual feature matches and then weighting and combining the individual scores to create a total score. By adjusting the weighting factors, it is possible to recognize objects with variable features.

CONNECTIVITY ANALYSIS

"Popular image processing" technique is connectivity analysis, so called because it breaks a binary image into its connected components. The connectivity analysis program builds a description of each blob (a connected component, either an object or a hole) as the image is processed. An array is created to hold information about the blob and its shape. Finally, a number of shape and size features values characterizing the blob are derived.[4,5]

The connectivity analysis can provide the following geometric information:

- Maximum and minimum values of width and height (X and Y values)
- Area
- Perimeter length
- Holes
- Centroid position
- Moments of inertia
- Orientation (from second moments)
- Elongation index = $\dfrac{\text{Major axis 2nd Moment}}{\text{Minor axis 2nd Moment}}$
- Compaction Index $-\dfrac{(\text{Perimeter})^2}{\text{Area}}$
- Linked list of coordinates on the perimeter

The connectivity analysis is performed in parallel with run-length coding. The processing time is directly proportional to the image area.

BINARY PROCESSING

The vision processing described previously, where pixel values are recorded as either "0" or "1" is referred to as binary processing. This technique is a "black and white" type of analysis. It can be used for geometric analysis and edge detection, but can not be utilized for analysis which require that surface characteristics be quantified.

GRAY SCALE IMAGE ANALYSIS

For advanced analysis, information may be required to aid complex part recognition or for the analysis of surface characteristics (i.e., texture, shade, pattern, etc). The gray level is a quantized measurement of image irradiance or brightness. The representation of the image as an array of brightness values is obtained from the digitizer. Various vision systems utilize different numbers of gray levels. For analysis by a 16-bit microprocessor, gray level scales are generally even digital power numbers: 4, 16, 64 or 256.

As an example of the use of a gray scale analysis, an inspection station may use a vision system in which the camera output is converted into 64 shades of gray by a digitizer. Each pixel would have a digital number assigned corresponding to the intensity it received. It may be specified that the part surface must register a gray scale reading of, say, 40±2 to pass the inspection. If the part brightness falls outside of the range of 38 to 42, the vision system may signal a robot to pick up that part and drop it into a reject bin.

A comparison of binary and gray level processing is presented in Figure 3-2 with two real time sampled images of SATYR.[6]

GRAY LEVEL HISTOGRAMS

In a gray scale analysis, gray level histograms may be constructed, as shown in Figure 3-3. The figure shows a histogram which represents the distribution of gray-levels in a picture where possible values range from 0 to 31. An algorithm can be constructed to search a histogram for peaks. The search is carried out by associating with each cell a value which depends on the height of the neighboring cells; if the cell is larger than both of its neighbors then it is a peak and is flagged as such; otherwise, the associated value is the address of the larger neighbor. When these values have been assigned, they are used as pointers in a search process which links together all the cells pointing to the same peak. These cells are all members of the same cluster. The weighted mean of all the gray-levels in the cluster is calculated and the scene is rewritten, replacing the original value of each pixel with the value of its cluster.

Template Matching refers to the use of a stored image in a computer. This technique is used after edge and region information is determined in order to ascertain position, dimensional inspection, etc.

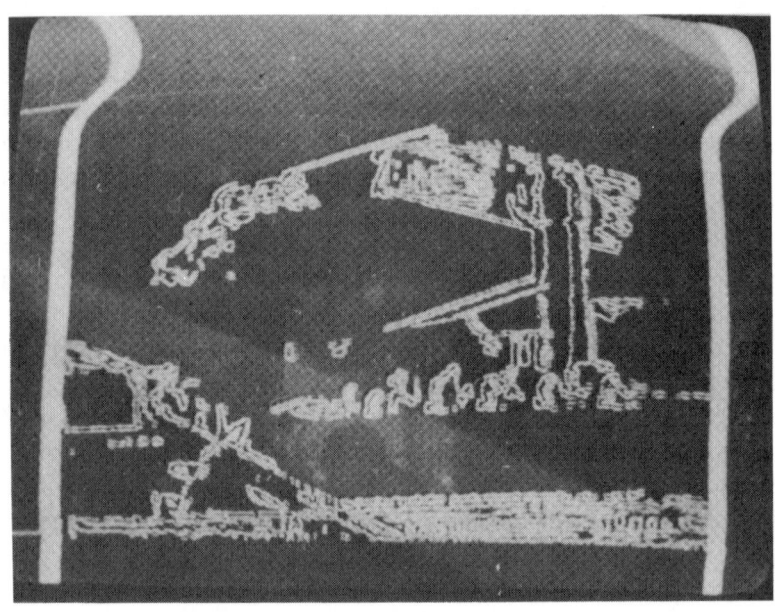

Figure 3-2. Comparison of binary processing and gray scale analysis on 'satyr' image.[6]

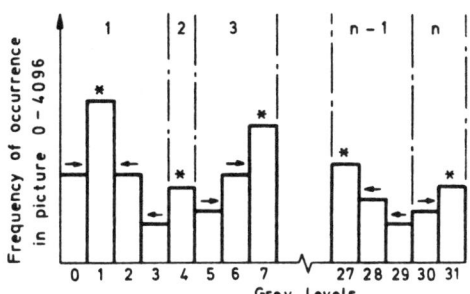

Figure 3-3. Gray level histogram.[19]

COMPUTER INTERFACE[7]

Many vision systems have their own language, as an example, VPL™ (General Electric Co.) RAIL™ (Automatix), BLIX (Machine Vision International). This means that for a user to integrate two or three pieces of equipment, he must learn two or three separate languages, and in many cases he may be reluctant to buy equipment from a manufacturer because his engineers and support personnel would have to be trained in yet another language. The proliferation of robot and vision languages is a major stumbling block to their universal integration.

COMPUTER ANALYSIS AIDS

Various computer techniques are used to improve analysis time, resolution or memory characteristics associated with the image analysis process. Some of these techniques are:

- *Frame Grabber:* A frame grabber stores in memory an entire frame of video information as a set of digital numbers corresponding to each pixel. This information is available to the computer for analysis.

- *Windowing*[8] : In windowing, the camera is used as a set of minicameras or a set of windows. A window is a portion of the image which is isolated and analyzed without regard to the rest of the image. A separate set of conditions can be set for each window, including the threshold. Each window can then be tested for a variety of features such as area, width, height, and location of specific edges. This type of inspection is very fast since processing is limited to the number of pixels in the window. A number of windows can be combined to provide the basis for a decision. The principle limitation of the windowing technique is that the part to be inspected requires orientation. Inspection of stampings for openings, and dimensions is usually done using windowing. The total inspection time is seldom more than one one second.

- *Run-length Encoding*[4-9] : In an effort to increase the speed of pattern processing and reduce the amount of memory required to store a large library of patterns, it is desirable to compress the data coming in from the camera so that only information about edges need be stored. A technique called run-length encoding provides this data compression. Analysis takes place concurrently with another process that actually recognizes patterns.

EXTRACTING EDGES[9]

A natural first step in analyzing a scene is to convert it into a sketch, that is, find the edges that separate regions of differing brightnesses. Edges correspond to abrupt changes in brightness. Such changes can be identified as places where the first derivative of the brightness is suddenly high or the second derivative is zero (see Figure 3-4). There are various schemes for doing this, all in some way related to taking brightness differences between adjacent points. The basic methods for extracting edge and line elements from images are[10-12]:

1. *Linear Matched Filtering*

    Successively convolve image windows with a template of the desired feature and seek the maximum value. Convolve means

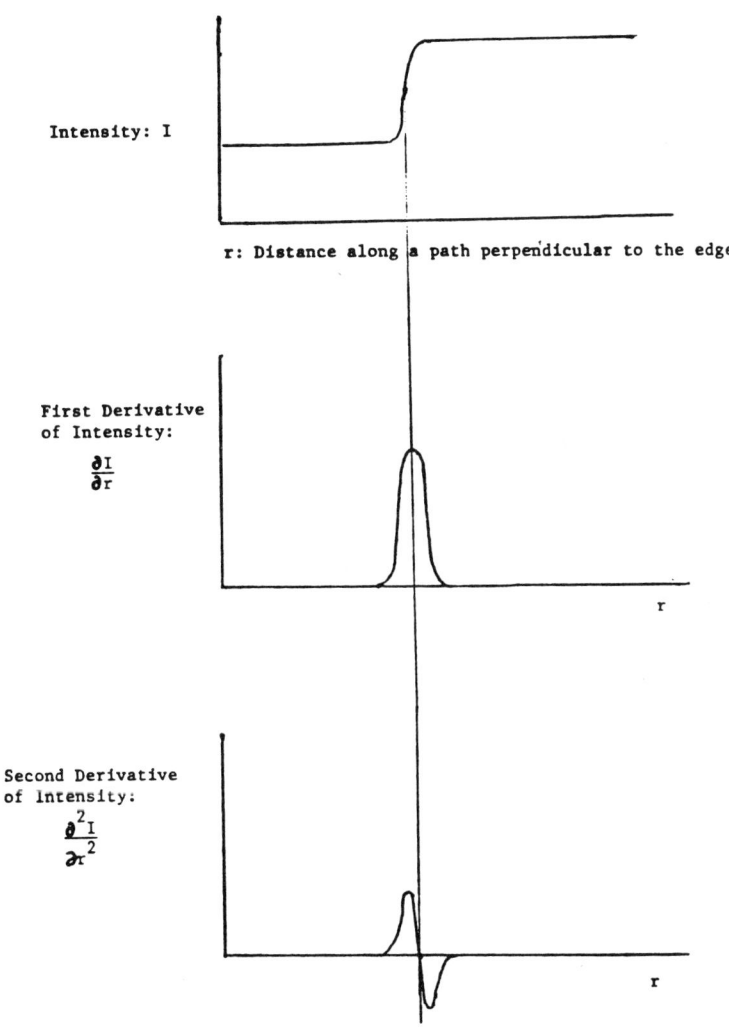

Figure 3-4. Intensity variations at step edges.[9]

superimposing an nxn operator over an nxn pixel area (window) in the image, multiplying corresponding points together and summing the result.

2. *Non-Linear Filtering*
Convolve windows in the image with a local operator (weighting function that approximates first or second derivatives by first or second differences). Examples of operators for doing this are shown in Table 3-1. In general, each point in the image is convolved with directional operators in as many directions as needed. The resultant outputs at each point are combined to determine the gradient vector (the orientation and magnitude of the intensity changes).

3. *Local Thresholding*
Apply local thresholding and discard responses that do not lie on boarders (between upper and lower threshold regions) and link responses that do.

4. *Surface Fitting — The Hueckel Operator*
Fit a surface to neighborhood of each pixel and compute maximum gradient of the surface. Consider as edge points those pixels having surface maximum gradients above a selected threshold value. This approach was first devised by Prewitt.[13] The Hueckel Operator is a popular method for doing this.

5. *Rotationally Insensitive Operators*
The Laplacian Operator (related to the magnitude of the derivative of the intensity gradient) is insensitive to the direction of a line and yields edge elements at pixel points where the Laplacian is zero. Thus discrete approximations to the Laplacian have proved useful in line finding.

6. *Line Following*
Shirai[14] devised a line following method that used a pair of parameters that varied according to how continuously and smoothly elements were found. These parameters determined thresholds for accepting a new element according to how close it was to the linear continuation of the current line being tracked.

MACHINE VISION SOFTWARE

Approach	Edge Criteria	Remarks
I. Edge Operators Detect first derivative of brightness: $\frac{\partial f}{\partial x}$ Sobel Operators $\begin{array}{\|c\|c\|c\|}\hline -1 & 0 & 1 \\ \hline -2 & 0 & 2 \\ \hline -1 & 0 & 1 \\ \hline \end{array}$ Edge Mask	pixels which yield max. values	operators tuned for limited range of operation
$\begin{array}{\|c\|c\|c\|c\|c\|}\hline -100 & -100 & 0 & 100 & 100 \\ \hline -100 & -100 & 0 & 100 & 100 \\ \hline -100 & -100 & 0 & 100 & 100 \\ \hline -100 & -100 & 0 & 100 & 100 \\ \hline -100 & -100 & 0 & 100 & 100 \\ \hline \end{array}$ Nevatia and Babu Operators	For each pixel, find angle operators that yield maximum value. Then thin, threshold and link (line fit).	similar operators used for each 30° angle.
II. Bar Operators Detect second derivative: $\frac{\partial^2 f}{\partial x^2}$ Bar Mask $\begin{array}{\|c\|c\|c\|}\hline -1 & 2 & -1 \\ \hline -1 & 2 & -1 \\ \hline -1 & 2 & -1 \\ \hline -1 & 2 & -1 \\ \hline \end{array}$	look for zero crossing	sensative to noise

Table 3-1
Examples of Non-Linear Filtering for Extracting Edge and Line Elements[9]

RANGE IMAGING SENSORS[15]

A range-imaging sensor measures the distances from itself to a raster of points in the scene. Although range sensors are used for navigation by some animals (e.g., the bat), hardly any work has been done so far to apply range image to control the path of a manipulator. This situation may change in the future however, as the technological and economical difficulties currently entailed with range-imaging sensors are overcome.

Different range-imaging sensors have been applied to scene analysis in various research laboratories. These sensors may be classified into two types, one based on the trigonometry of triangulation and the other based on the time of flight of light (or sound).

Triangulation range sensors are further classified into two schemes, one based on a stero pair of television cameras (or one camera in two locations), and the other based on projection of a sheet of light by a scanning transmitter and recording the image of the reflected light by a television camera. Alternatively, the second scheme may transmit a light by a rocking receiver. The first scheme suffers from the difficult problem of finding corresponding points in the two images of the scene. Both schemes have two main drawbacks: missing data for points seen by the transmitter but not by the receiver and vice versa, and poor accuracy for points that are far.

The above drawbacks are eliminated by the second type of range-imaging sensor using a laser scanner, which is also classified into two schemes: one based on transmitting a laser pulse and measuring the arrival time of the reflected signal, and the other based on transmitting amplitude modulated laser beam and measuring the phase shift of the reflected signal. A simplified block diagram of the latter sensor is shown in Figure 3-5. The transmitted beam and the received light are essentially coaxial.

Range-imaging sensors have been applied so far primarily to object recognition. However, they are also very suitable for other tasks, such as finding a factory floor or a road, detecting obstacles and pits, and inspecting the completeness of subassemblies.

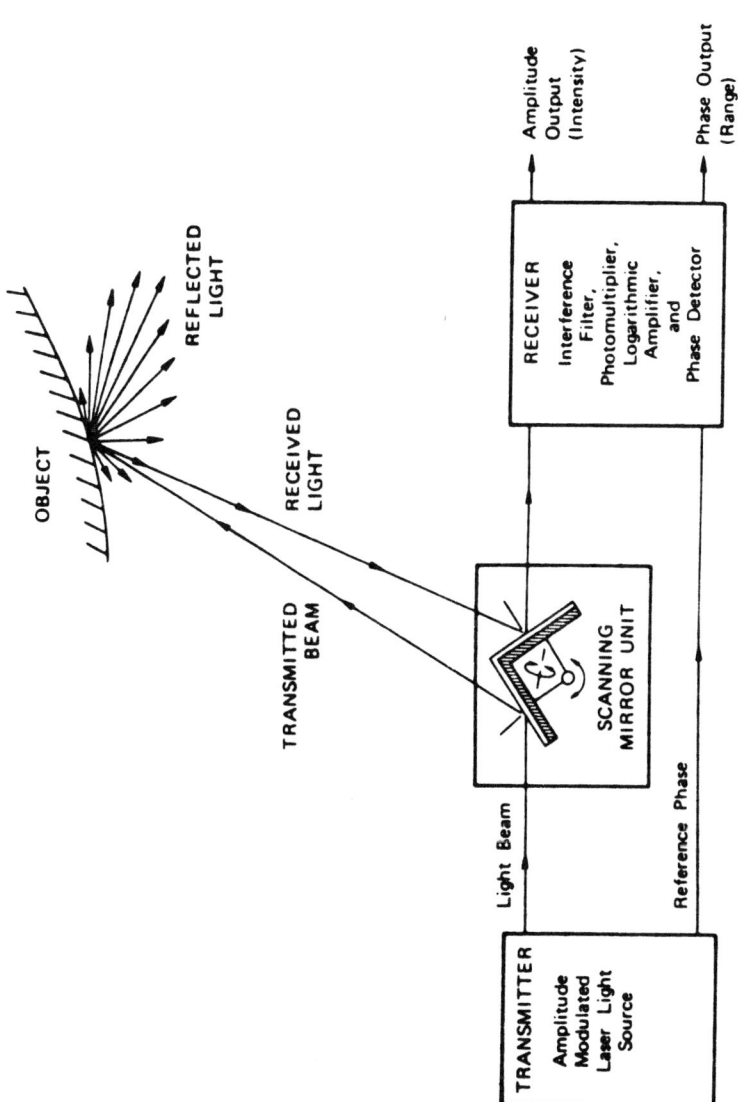

Figure 3-5. Simplified block diagram of a range imaging sensor.[15]

STANFORD ACRONYM SYSTEM[16,17]

A new "Acronym" geometric modeling/reasoning system under development at Stanford University will model real-world industrial parts (both specific objects and generic classes of parts), perform programmable inspections, pick parts from bins, and perform automated assembly operations, all by extracting 3-dimensional information from monocular images. To provide general-purpose vision for manufacturing, CAD-generated models help predict observable image features by employing generalized cones defined by a planar cross-section and a space-curve spine, through use of a "sweeping-rule" concept.

In its implementation, Acronym applies ribbons (planar shapes—2-dimensional specializations of generalized cones—described by 3 components) and ellipses (to describe shapes generated by ends of the generalized cones). If an object is modelled with a well-determined size, position, and orientation, then combining constraints from hypothesized matches for many objects will make that object useful in determining parameters of the camera and other objects. By this procedure a mobile robot can use known reference objects as a means of determining its own absolute location and orientation as well as defining those parameters of other movable objects. In this way, in a bin-picking task, a manipulator can be commanded to pick up an object from the bin.

The vision system involved in this work relies upon propagation of symbolic algebraic constraints for predicting image features as well as for extracting 3-D information from image-feature measurements. This research is being integrated with extensive studies to find image-intensity boundaries in images and to describe surfaces through stereoscopic vision.

An example of the Acronym system is shown in Figure 3-6.

COLOR SENSING

Color sensing can be useful for industrial application such as assembly, painting and inspection for surface damage. Many assembly components are color coded, and color analysis would provide a logical approach in the automation of such operations. At present, the use of color analysis for machine vision in industry is still un-

MACHINE VISION SOFTWARE

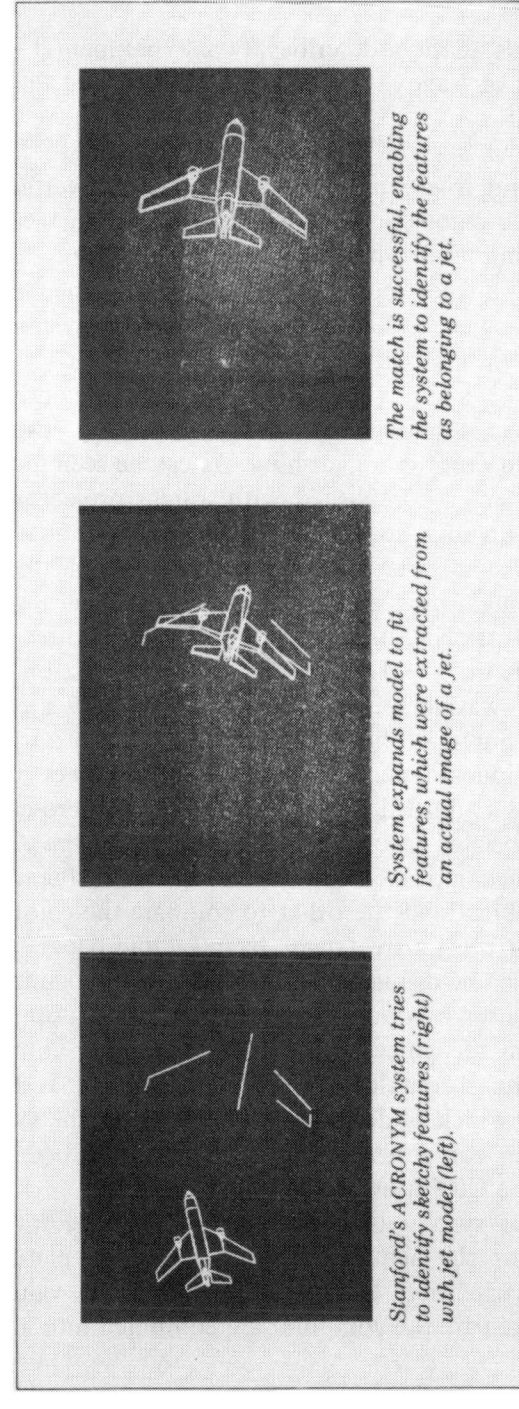

Figure 3-6. Example of Stanford's ACRYNYM vision system.[1]

Stanford's ACRONYM system tries to identify sketchy features (right) with jet model (left).

System expands model to fit features, which were extracted from an actual image of a jet.

The match is successful, enabling the system to identify the features as belonging to a jet.

developed. This section describes two experimental developments by researchers at Nagoya University[18] (Japan) and Philips Research Laboratories.[19]

The Nagoya System[18] consists of three photo diodes with a color filter, a micro-computer, and A/D, D/A converter. This system can know three color components of an object surface and instruct the robot using the robot language. This system and the robot control system are connected to each other with the GP-IB interface. The robot can be controlled by this color sensing system, so the total system has much flexibility.

The color sensing system is shown schematically in Figure 3-7. The most basic functions of the color sensing system are color recognizing and color distinguishing. As can be seen in the schematic diagram, they are accomplished with a light projector, three photo diodes with individual color filter, three inputs of an A/D converter, a single board computer with programs in EPROM for color recognizing and distinguishing, and a CRT which can receive the commands from an operator. The dimension of an array of three photo diodes is shown in Figure 3-8. The filters whose transmission coefficient as a function of wavelength is shown in Figure 3-9 are mounted on three isolated SnO_2-Si photo diodes. And so each output voltage of three photo diodes depends on color of the object paper. Each of these signals is amplified and received by the micro-computer "80/30-G after translated into digital signals through the A/D converter. The 80/30-G accomplishes two functions, one of them to decide the color of the object and the other to compare the color of the object with the color previously taught by a sample paper. The results of color sensing can be displayed on the CRT as, for example "GREEN" and/or can be used by other I/O equipment using a simple command.

The system recognizes color under computer command. Figure 3-10 shows the thresholds for color recognizing. If R', B', G' are in each range which is defined in Figure 3-10, the system displays its color name, components ratio and output voltage of each color amplifier on the CRT using RS232C interface.

The Philips system uses the hardward configuration shown in Figure 3-11 for the acquisition and display of pictures. Color is recognized as triple red, green and blue (R, G, B) values. The three pictures acquired by the computer are combined into a single picture

MACHINE VISION SOFTWARE

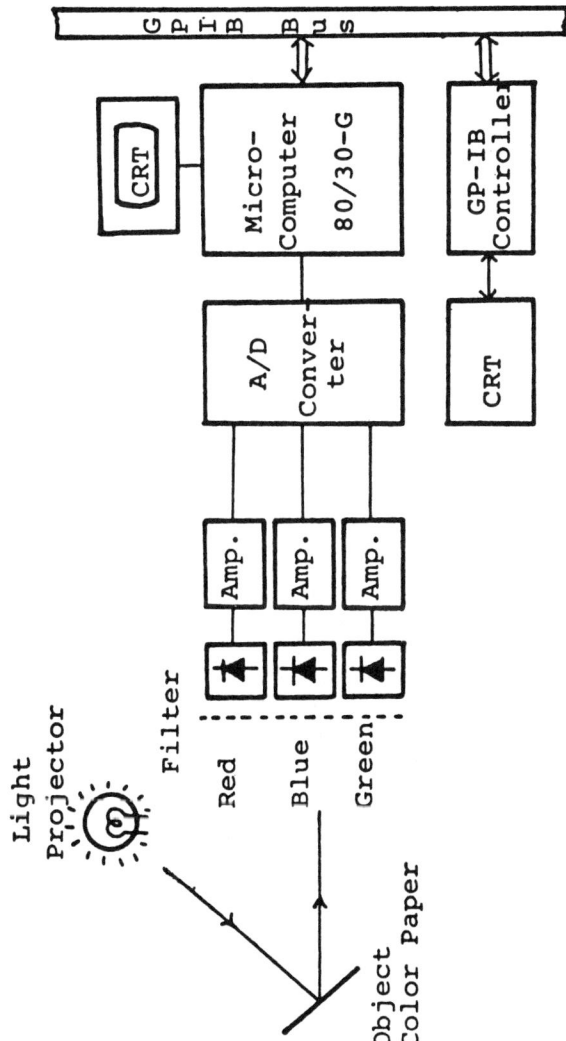

Figure 3-7. A schematic of the Nagoya color sensing system.[18]

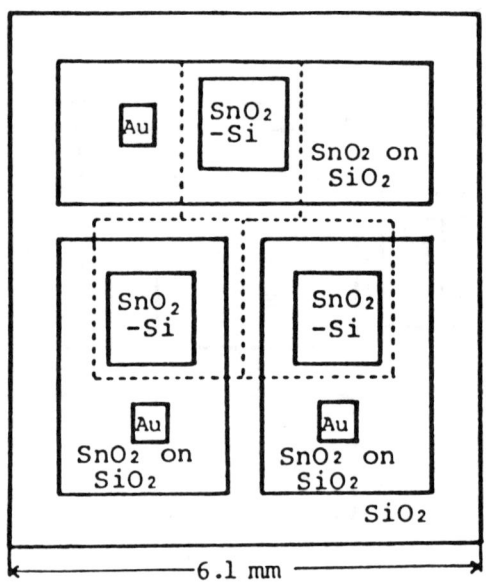

Figure 3-8. Diagram of photo diodes. Broken lines show a filter composed of three primary colors.[18]

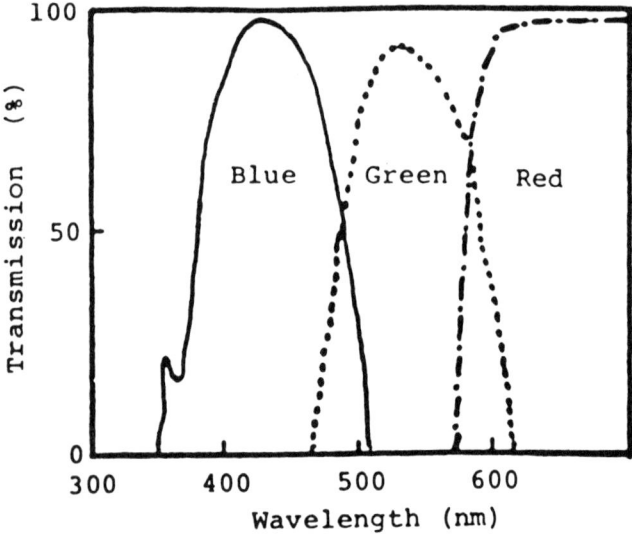

Figure 3-9. Transmission coefficient against wavelength of blue, green, and red filters. [18]

MACHINE VISION SOFTWARE 41

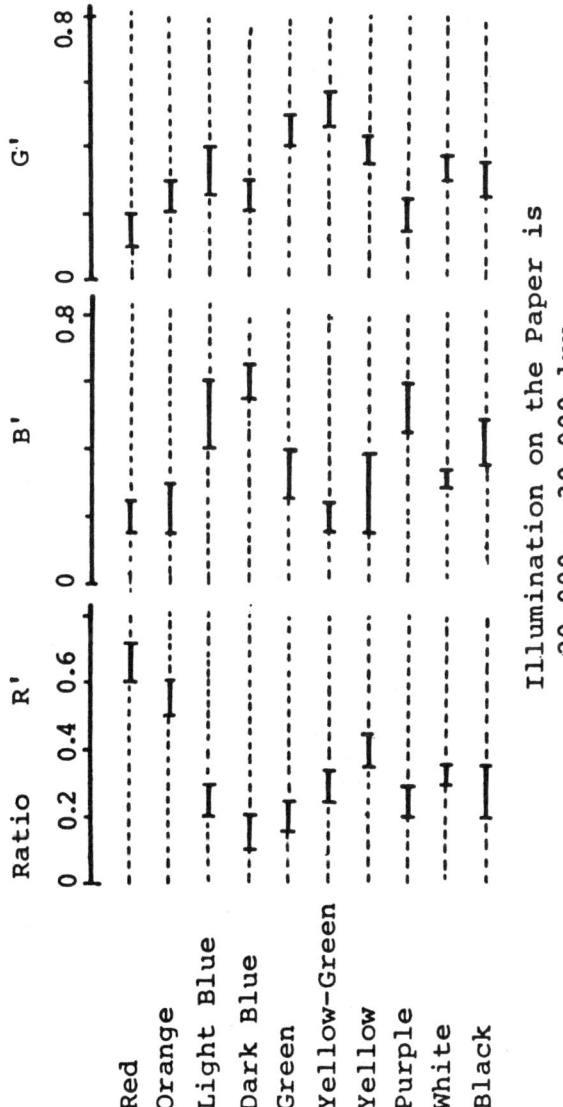

Figure 3-10. Experimental results of color processing.[18]

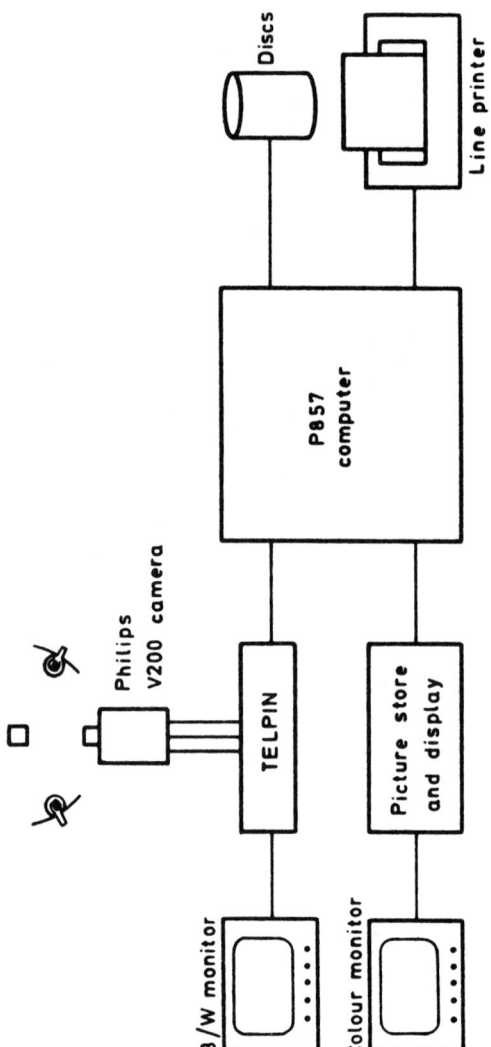

Figure 3-11. Hardward configuration for Philips color vision processing system.[19]

MACHINE VISION SOFTWARE

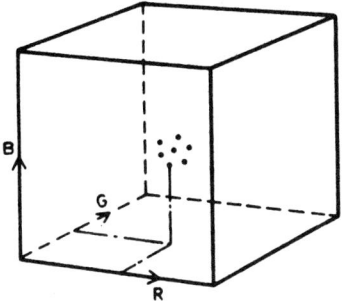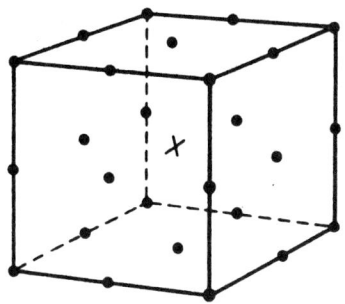

Figure 3-12. Diagram of pixel cluster in color space (left) and showing 26 neighbors in color space (right).[19]

in which each color is represented by a unique integer value. This is accomplished by regarding each color as consisting of three co-ordinates in color space (see Figure 3-12). The co-ordinates are packed, by means of logical operations, into a single computer word. The color resolution is four bits per pixel, giving a packed word length of twelve bits and a color range from 0 to 4095. This implies that different colors may have the same hue but different brightnesses.

Any color in a scene is represented by a point in color space and similar colors are represented by neighboring points. These are called group clusters. In comparison with a gray level histogram (see Figure 3-3), a color histogram contains 4096 values rather than 32 and each point in color space has, not two, but twenty-six neighbours (Figure 3-12).

FUTURE TECHNIQUES

Some future techniques which may be expected in machine vision have been listed by Gevarter[20]:

1. Greater use of direct range measurements—e.g., scanning laser radar, or LED's and photo diodes using phase or intensity measurements.
2. Parallel processing to obtain feature extraction in real-time.

3. Improved methods for determining edges and regions.
4. Use of optical markings for recognition, for optical code reading or for determination of pose.
5. Use of intrinsic image analysis—to determine surface orientation, depth, relationship of foreground to background, and texture.
6. Scene analysis for less-structured situations.

References

[1] Kinnucan, Paul, "How Smart Robots Are Becoming Smarter," *High Technology*, v.1, n.1, Sept/Oct 1981, pp 32-40.

[2] Agin, G.J., "Computer Vision Systems for Industrial Inspection and Assembly," *Computer*, May 1980.

[3] Nitizan, D., Bernard, S., Bolles, R., et al., "Machine Intelligence Research Applied To Industrial Automation—Tenth Report," NSF, 1980.

[4] Trombly, John, "Adding Machine Vision To Assembly Lines," *Machine Design*, November 11, 1982, pg. 78-81.

[5] Rosen, C.A. and Gleason, G.J., "Evaluating Vision System Performance," *Robotics Today*, Fall 1981.

[6] Barthes, J.P.A. and Zavidovique, B., "How Much Intelligence Should We Expect From A Vision Processor In A Multi-Processor Robot System?," Proceedings of the First International Conference on Vision and Sensors," IPC Publications, U.K., 1981, pp 161-167.

[7] Robertson, Gordon I., "Heirarchial Control of Intelligent Robot and Vision Allows Plug-In System Integration," Proceedings of AUTOFACT IV, SME, 1982, pp. 11-35 to 11-50.

[8] Nowak, Glenn, "The Advent of Machine Vision Systems," *Manufacturing Engineering*, November 1982, pg. 56-60.

[9] Gevarter, William B., "An Overview of Computer Vision," NBS report no. NBSIR 82-2582, September 1982.

[10] Rosenfield, A., "Image Pattern Recognition," *Proceedings of the IEEE*, Vol. 69, No. 5, May 1981, pp. 596-605.

[11] Gennery, et al., "Computer Vision," JPL Publ. 81-92, Nov. 1, 1981.

[12] Brady, M., "Computational Approaches to Image Understanding," M.I.T.A.I. Memo No. 653, October 1981.

[13] Prewitt, J.M.S., "Object Enhancement and Extraction," in *Picture Processing and Psychopictories* (B.S. Lipkin and A. Rosenfeld, eds.) New York: Academic Press, 1970, pp. 74-149.

[14] Shirai, Y., "Analyzing Intensity Arrays Using Knowledge About Scenes," in Winston, P.H. (ed.), *The Psychology of Computer Vision*, McGraw-Hill, 1975, pp. 93-114.

[15] Rosen, Charles A. and Nitzan, David, "Use of Sensors In Programmable Automation," *Computer*, vol. 10, no. 12, pp. 12-23.

[16] Brooks, R.A. and Binford, T.O., "Geometric Modeling in Vision for Manufacturing," Proceedings, Society of Photo-Optical Instrumentation Engineers, Vol. 281, Techniques and Applications of Image Understanding, pp. 141-159, 1981.

[17] Industrial Robots International/August 9, 1981, pg. 6.

[18] Urda, M., Matsuda, F. and Sako S., "Color Sensing System For An Industrial Robot," Proceedings on the 10th International Symposium on Industrial Robots, 1980, pp. 153-162.

[19] Connah, D.M. and Fishbourne, C.A., "The Use of Color Information In Industrial Scene Analysis," Proceedings of The First Conference on Robot Vision and Sensory Controls, IFS Publications, Ltd., U.K., 1981.

[20] Gevarter, William B., "An Overview of Artificial Intelligence and Robotics, Volume II — Robotics," Natural Bureau of Standards, NBSIR82-2478, March 1982.

4
Checklist for Machine Vision Applications

Machine vision is a much reviewed subject in the trade periodicals of virtually every industry today. In metalworking, not a month goes by that one or more articles describing an application are not published.

Machine vision is still not a well defined entity. If one considers capital equipment that is a substitute for a human visual function, then machine vision includes substantially more technology that is on the market today. For example, the products that use laser techniques to perform gaging to control wire dimensions during drawing can be considered machine vision. Products that perform off-line dimensional measurements automatically are forms of machine vision.

In the purist sense machine vision is coming to mean manufacturing systems that operate on image data. Today companies offer machine vision products which are multi-functional for end-user application (configurable vision systems that are "solutions looking for problems"). Their configurability is restricted to a certain performance envelope. Significantly the artificially intelligent eye is not on the market yet.

Other companies offer products which have a specific performance envelope designed for a specific task, such as the off-line dimensional measurements mentioned above. Virtually any size/shape part can be measured so there is indeed flexibility built into the design of these systems. However, gaging is the only function these systems can perform. They can not inspect for blemishes, for example.

Still other companies exist that are prepared to integrate complementary techniques. The result becomes an inspection station analogous to fixed automation, dedicated to the inspection of one object. In most cases, the actual inspection heads are "off-the-shelf" though in the heads themselves one has a certain element of flexibility. That is, they can be configured to address the inspection of many objects, but once configured for an object their capabilities are fixed.

There are some technological barriers when it comes to processing image data. Today these include:

1. The resolution of the sensors.
2. The time to form an image.
3. Typically only examining the two dimensional projected image of a three-dimensional image.
4. The ability to process color.
5. The time to process an image to make a decision.

People enjoy a fantastic advantage since they have an uncanny ability to resolve detail in three-dimensional color scenes when that detail represents an exception to a norm. Furthermore, people naturally process scene data in parallel.

Today imaging sensors have an ability to essentially dissect a scene into an array of 300 x 300 picture points (pixels) in 1/30th of a second. There are cameras that can dissect a scene into 2000 x 2000 pixels but they take 1.5-2 seconds to do so. Fortunately the technology is advancing and by the end of 1987 there should be cameras on the market capable of dissecting a scene into 1000 x 1000 pixels in 1/30th of a second.

An alternative way of obtaining high resolution image data is by building up a picture based on a series of digitized lines. This is the basis of line scanners. These have the ability to resolve a line into up to 4096 pixels at a maximum rate (given by one manufacturer as) $N \times 0.67 \times (10)^{-6}$ second, where N is the number of diodes (pixels) in an array. As with a photographic camera, the longer the exposure time (line scan time) the less the light intensity requirement to produce a signal from the image. On the other hand, the shorter the exposure time the less the image is blurred by motion of the object being scanned.

This brings up the subject of motion. If the object to be examined is in motion, capturing a picture requires the use of a synchronized strobe, if one is to use a matrix array camera. Both with the use of strobe and a line scan camera one still has to pay attention to the affect of smearing images in the process of capturing the image.

Machine vision systems today have an ability to interpret shades of grey but not color. In other words, within a given scene all the same saturation levels/luminous intensities in the scene regardless of color will be characterized as the same shade of grey. This is compounded somewhat because various sensors have different spectral responses over the visible spectrum of interest in most vision applications.

Processing color requires the use of filters or the use of a recently introduced "black box" from Synthetic Vision Systems which permits the selective integration of red, blue or green data serially from a color television camera. So while technically possible the amount of data to be processed goes up by a corresponding factor.

Another shortcoming of today's machine vision systems is that they only analyze the two-dimensional projected image of a scene. The detail they "see" is analogous to that captured in a photograph. Consequently, when reject conditions may be obscured by the shape of an object, more than one camera is required to view an object from different angles. This costs time in processing and merging the independent scene data from each of the cameras.

The image processing/decision making speed is today limited by computer technology. Companies that offer parallel-pipeline image processing architecture (such as Applied Intelligent Systems, Machine Vision International, and Synthetic Vision Systems) claim the highest data processing rates, up to 3 BIPs.

Some companies have developed hardware co-processor techniques and, where applicable, have the ability to also handle on the order of up to 40 MIPs. Recently companies with links to the semiconductor business have announced the development of special integrated circuit chips that perform some image processing functions at rates up to 100 MIPs. Because of these limitations, technology other than machine vision (better characterized as remote sensing) is also reviewed in this report where applicable.

As a point in passing, it has been estimated that people process image data at a rate of 30 to 100 BILLION instructions per second.

The challenge with people is attentiveness for extended periods of time. Studies have shown that for applications such as those involving the sorting of bad product from good, based on visual appearances, people who operate in crews which alternate tasks every 20-30 minutes (so inspection represents a function they perform one-third of the time) are only about 85% effective. That is 15% of the "bad" product will pass. As can be appreciated, even with two inspectors operating serially each fully attentive, 2.25% of the "bad" product will pass. Significantly, studies have also shown that people are also prone to reject falsely good product.

There are many things to look for when reviewing your facility for potential machine vision applications. An installation makes sense to detect a reject state at the point of lowest value added. By incorporating statistical process control techniques a system can spot trends indicative of pending out-of-specification conditions. An ancillary benefit is reduced paperwork since record keeping is automated.

Rejects can be separated into scrap that can be reclaimed from that which can not. Another area to investigate is one where expensive hard tooling is required to hold a part for an operation. This may be avoidable all together or at least cheaper flexible fixturing substituted if a machine vision system is used. In this case the system can provide location analysis—"software fixturing" so to speak. A key to this type of requirement is where set-up time is lengthy and the amount of time a part is actually being operated on is very small relative to the total cycle time associated with an operation.

If a high incidence of equipment breakdown is being experienced because of such problems as over/undersize or misshapened/warped parts, a machine vision system upstream of the feeder mechanism can reduce or even eliminate downtime.

A situation that definitely warrants a machine vision system is one that involves inventorying parts because inspection may result in the rejection of a complete batch based on statistical sampling techniques. 100% inspection assures every part is good so "just-in-time" inventory can be a by-product, with a corresponding reduction in material handling and damage that might be experienced by handling. Similarly, machine vision opportunities exist where inspection is a production bottleneck.

As with the justification for robots one can look for applications related to unhealthy or hazardous environments. Possibly OSHA is expressing concern about the operator's well-being—the noise level is too high, the temperature too hot, products are too heavy, etc. It may be the environment includes contaminants (metal dust or vapors) which can be injurious to a person. The converse may also be a justification—people bring contaminants into the environment which can damage the product—dust causing damage to polished surfaces, for example.

Where an operation experiences errors due to operator judgement, fatigue, inattentiveness or oversight brought about because of the dullness of the job, machine vision opportunities exist. Certainly when an operation is experiencing a capital expansion mode, machine vision should be considered in lieu of alternative less effective more costly methods.

Unquestionably any operation can identify opportunities for machine vision by performing an introspective examination of its operations. The adoption of this technology with the result of objective 100% inspection of products will cut costs, improve quality, reduce in warrantee repairs, reduce liability claims, and improve consumer satisfaction—all components in an improved profit picture.

BEFORE UNDERTAKING A MACHINE VISION PROJECT KNOW YOUR COMPANY

There are many factors that should be taken into consideration before proceeding with a machine vision installation. Fundamental factors include recognizing the short- and long-term manufacturing philosophy of the company. For example, is there already in place or under consideration the wherewithal to tie together the manufacturing process via a hierarchy of controllers and computers? Should this be anticipated? This, therefore, dictates the use of a machine vision system with compatibility—the ability to be interfaced to and communicate with an arrangement of computers. Essentially then the machine vision system becomes a computer peripheral.

This is an important consideration when one must accommodate the inspection of products manufactured in small batches. If no means is available to down load inspection programs then the system will have to be retrained at the beginning of each run. The result could be a significant set up time which will interfere with efficient batch production. Even if provision is made for local program storage on a cassette or floppy disc, are there so many models to be concerned with that a second system will be required to be used for "training"– the development of the "golden" files. Interfaceability back to a CAD data base which dictates the inspection criteria on the basis of design rules or its own "golden" file will be far more efficient.

Other considerations include: is the contemplated installation for productivity improvement or to expand production capacity? In either case is the machine vision system to be delivered a "stand alone" system or is the installation to be "turnkey?" The former case implies ultimate responsibility for making the system work rests with the buyer. This in turn implies the buyer must be prepared to train staff to become reasonably familiar with image processing theory as well as system properties so he can optimize the performance of the system.

As a "turn-key" the machine vision supplier or a systems house assumes total responsibility for making it work. The end-user never has to understand why it works—simply that it does work. In this case the supplier must become familiar with the buyer's manufacturing process. For example, it may appear that an operator is simply performing a sorting function, separating containers by their size or shape. On closer scrutiny one will also observe that once in a while the operator throws a container into a scrap bin. While probably not classified as an inspector, by virtue of innate intelligence the operator knows enough to separate incompleted or misshapened containers.

By virtue of increased sensitivity to exceptions, an operator can become more sensitive to specific conditions. For example, because training has emphasized the importance of separating all green containers, the person has a heightened awareness when such objects pass. Similarly, a machine vision system might have to somehow include a weighting factor (a complementary color detector, for example) that will increase sensitivity to a specific factor—color in this example.

The converse is also true—an operator may be trained to be tolerant of color shade variations; for example, all yellows regardless

CHECKLIST FOR MACHINE VISION APPLICATIONS

of pale gold or virtually orange might be acceptable since it's the basic color itself that provides the distinction. In this case, therefore, the machine vision system must be equally tolerant of these normal variations while also maintaining sensitivity to the fundamental defects the system is supposed to capture.

In other words, a successful "turn-key" installation requires the buyer develop a specification that correctly establishes the criteria characterizing a reject condition. The development of the specification may also provide focus to the type machine vision technology that will be required for the installation. For example, if color variation must be tolerated and only shape monitored, a system which does not operate on shades of grey may be more appropriate. Similarly, a backlighted arrangement might be more appropriate so only silhouetted properties are captured and operated on.

Other "systemic" considerations might include: how old is the capital equipment and the manufacturing process itself? Does it make sense to augment the capabilities of the equipment if it has already been fully depreciated and may be replaced in a year or two because of technological changes in materials, for example? Are there manufacturing technology breakthroughs that may be taking place that will result in wholesale replacement of capital equipment? Keyboards represent a good example. For years the characters on the key caps were developed by injection molding techniques. Now the keyboard industry has largely adopted a transfer printing process which in one shot transfers all the legends onto the key caps. The inspection problem is completely different. Whereas before, key transposition was a major cause for rejection and legend quality only of secondary concern, with the printing technique, transposition problems are all but eliminated but legend quality is more difficult to control.

Other managerial philosophies must be examined. Is there a "build or buy" decision contemplated? That is, is there a possibility the company may seek outside vendors in the future rather than produce it internally? Is the emphasis of management being placed on making it right in the first place and, therefore, monitoring the production process to avoid rejects as opposed to culling rejects in final inspection?

Are there risk-takers in the organization willing to stick their necks out to change a situation or is it strictly a "laissez-faire" organizational philosophy that prevails? "We beat the Germans in WWII without robots in the battlefields so why do we need them now?" Is there a management concern for the employee which is the motivation for considering machine vision automation—his health—avoiding hazardous or hot environments?

All these factors play a role in specifying a system. Similarly, understanding these factors beforehand can make the difference between a "white elephant" or a successful installation. The message should be clear—know the company before proceeding with the identification and feasibility assessment of a machine vision installation.

EXAMINING NEEDS

Having assessed the general philosophy of management, one can now systematically examine needs on a department-by-department basis. There are opportunities in every department: in-coming receiving, forming operations, assembly, testing, work-in-process monitoring, and warehousing. Significantly, these opportunities may not be for machine vision systems in the purist sense of the meaning of machine vision where the term connotes the processing of image data.

There are on the market many products today designed to substitute for a human visual task which may not necessarily satisfy the conventional definition of machine vision. That should not deter the technique from being considered.

In any event, things to look for include:

— How are goods routed for the vision inspection task presently?
— Does the operation require automating loading/unloading?
— Is the operation inventoried?
— Are products stored in bios? magazines? etc.?
— What are the actual inspection functions that must be performed by the system?
— Gaging—is it now being done by "eyeball" or with instruments (micrometer)?

- Cosmetic inspection—is it a detail inspection or is it now performed by a cursory look for more gross appearance differences?
- Does the inspection first require identification? If so is it by shape? by reading characters, etc.?
- Is the inspection itself one of verifying shape conformance? Again, with or without instruments?
- Is the operation one of just verifying that an assembly is complete before another operation is performed or that objects are oriented properly, etc.?
- If, in fact, it is a complete assembly operation that is performed, does the operator locate parts and guide them into place?
- Is this operation a combination of several of the above tasks—location guidance, cosmetic examination and gaging, for example?
- Is present task 100% inspection or sample inspection?
- Does the operator perform tasks other than vision? Assembly, machine loading, etc. for example?

When a machine vision task has been identified, scoping out the application involves a detail examination of the task being performed. Input should be solicited from those presently performing the task not just by observing those performing the task. Things to look for in the application include:

- Scene complexity—is there contrast in what must be observed, i.e., can you visually see the condition without picking up the part to manipulate it in the light?
- How small is the smallest detail you want to see? How big is the object/field-of-view? As in photography one can see both large and small objects with television cameras, but the detail that can be detected reliably is proportional to the field-of-view. Where necessary machine vision systems can employ more than one camera so detail versus field-of-view need not be a factor which would preclude considering machine vision.

- Are there normal variations in the appearance of the object that are ignored? A vision system will somehow have to normalize those conditions.
- Is the part repeatedly and consistently located? If not then image capture will dictate a requirement for an even larger field-of-view, reducing the detectable detail. It will also require location analysis to reconcile the image captured with respect to position.
- Are parts overlapping? touching? jumbled? Can they be presented registered and not touching?
- Are the parts moving or indexed where they can be held in place in front of the scanner? If in motion at what conveyor speed? How well regulated is the speed?

Other things to consider include:

- Does the operator perform three-dimensional analysis?
- Is the decision based on color interpretation?
- How much time is there to make a decision?
- What of the performance of any system substituted for a person? What percent of the reject objects will you tolerate pass as good? What false reject rate (number of good units that are rejected) will you allow? No system is perfect!
- How much start-up time will you allow? If the system is not dedicated forever to a specific task, how much time will be permitted to get the system ready between product change-overs?
- How much floor space is available? Overhead space that may be needed to mount cameras/lighting arrangements? Will much equipment rearrangement be required? Is power, air, etc. available? How much downtime can you tolerate for other than routine maintenance?
- In what kind of environment will the equipment operate? Dirt/dust? grease/lubricants? water? shock and vibration? temperature/humidity? electrical noise?
- To support the specification of a system can the following be made available to the prospective vendors for bidding: job

descriptions, present specifications, drawings, samples, photographs/videotapes of the facility and inspection area? What kind of personnel will be available to operate and service the system following installation?

Now that you have given consideration to all the above, you are ready to go forward with the implementation phase. Successful implementation requires the involvement of all those to be affected by the change. Communications is required—of management's objectives, of everyone's responsibilities, or progress.

Having conducted a preliminary needs analysis, a preliminary feasibility analysis is required. A preliminary specification should be written, timetable established, and 5-6 prospective vendors identified. The vendors should be informally solicited at this time for budgetary estimates. These should then be used to conduct a preliminary cost analysis. If all is reasonable, a more serious bid cycle should begin.

A formal RFP should be written that includes a description of the project in detail along with a brief discussion of the business of the operations. The rationale for the project solicitation should be reviewed as well as the schedule. The functional requirements should be spelled out in detail—where possible separate the "needs" from the "wants." The RFP should also define the response wanted: schedule, training, service, warranty, documentation, installation support.

Along with the RFP, representative samples should be forwarded to each of the solicited vendors. If set-up costs are involved, a fee should be anticipated to cover those costs. The samples should be representative of all production and include good and bad samples as well as samples that are marginally good and marginally bad.

A vendor conference should be planned to be conducted at the prospective installation site 10-14 days after the RFP is received.

The vendor's conference does a number of things:
- it is an opportunity to clarify objectives and requirements
- demonstrates the vendor's interest
- permits a facility review and first-hand appraisal of the operation on the part of the vendor

- being a one-time session, it avoids the need to tie up staff for individual visits by each of the vendors.

Finalizing the evaluation should include visits to responsive vendors to make sure there is complete understanding. Where possible, referenceable vendor installations should be contacted and visited, especially where there are parallels with the application presently planned.

Reach a decision on the most qualified vendor based on:
- previous successes (especially related successes)
- quality of work
- reputation
- ability to meet schedule
- understanding of your business application.

At this time you are in a position to determine the total project cost and estimate ROI. Included in your estimate of the system cost should be:
- acquisition cost
- hardware/software
- custom content
- site preparation
- training
- installation
- maintenance
- consultant

plus operational costs such as personnel, maintenance, supplies (air, printer paper, etc.).

Having justified the expenditure it is time to issue a purchase order. This should include: all documentation, schedule, method to resolve disputes, specifics on acceptance criteria and acceptance testing, and training. Milestones should be worked out with the vendor and, when warranted because of the complexity of the installation, meetings with the vendor should be regularly scheduled around benchmarks.

CHECKLIST FOR MACHINE VISION APPLICATIONS

A systematic procedure in the evaluation of needs, the development of specifications, the planning for a project, and the procurement and installation of a system will virtually guarantee a system that meets the requirements and is delivered on schedule.

5

Industrial Applications for Inspection

It has been estimated that 90% of all industrial inspection activities requiring vision will be done with computer vision systems within the next decade.[1] Machine vision for inspection, however, is not just a technology for the future. Immediate, practical applications for automated inspection devices in U.S. industry are estimated in the hundreds of thousands.[2] General Motors alone has suggested they have 44,000 requirements.

This chapter presents automated inspection applications of machine vision which have been demonstrated based on information provided by manufacturers and published literature.

CONTAINER AND LABEL INSPECTION

One of the most common applications for vision systems is for the inspection of containers (bottles and cans). The containers may be inspected for a variety of potential irregularities:

- missing, crooked or incorrect labels
- bottle fill height
- missing or damaged caps
- foreign can detection
- code dates

A number of companies have product offerings that specifically address these requirements of the packaging industry. Most, however, are limited in what they can perform because of: registration compensation difficulties (especially with round containers), resolution limitations of the sensors, contrast difficulties that do not set label off from container adequately, detail that must be detected (some suggest need to proof-read labels), and acceptable variations in hue saturations and brightness in both the hues on the label and container.

DIMENSIONAL INSPECTION OF COMPLEX PARTS

The ability to make dimentional measurements is limited by the contrast in the scene that sets edges off. Because contrast is often low, systems that perform dimensional analysis have an ability to operate on edge gradients to do sub-pixel analysis—detect an edge location to a fraction of an inch (claims to 1/64th of a pixel are made in certain applications).

Regardless of sub-pixel processing capability, another challenge relates to the limitation of the sensor's ability to see detail. Brute force analysis suggests that the smallest detail a camera with a 512 x 512 array can detect is 2 mils on a side. Consequently, arrangements of cameras, arranged in a fixed relationship to each other, are often used to make measurements on large objects. In this case, the edge detected in each of the cameras is related to a reference location and the differences analyzed and compared to the tolerance for the measurement between the two edges.

The difficulties associated with the inspection of complex parts can be appreciated in the illustration of Figure 5-1. One part has a missing tab. However, such a defect may not be noticed by a human operator. In this example, the defective part is identified by a Control Automation, Inc., CAV-1000 vision monitor.

Metal stampings are subject to cracks, breaks and tears in certain areas due to the stresses induced by the drawing operation. Many stampings are quite complex, making operator inspection time-consuming and complex. An example of such a part is an automobile dash panel, containing 50 holes and clinchnuts. Figure 5-2 shows this part undergoing automated inspection by the Opto-Sense Vision System of Copperweld Robotics. The system utilizes four cameras which are located in an enclosure box to eliminate extraneous light sources.

A 3-dimensional vision inspection system was developed by Robotic Vision Systems, Inc., Melville, N.Y. An installation of the system at Cummins Engine Company for the inspection of engine castings is shown in Figure 5-3. Over 1250 points are checked in

INDUSTRIAL APPLICATIONS FOR INSPECTION 63

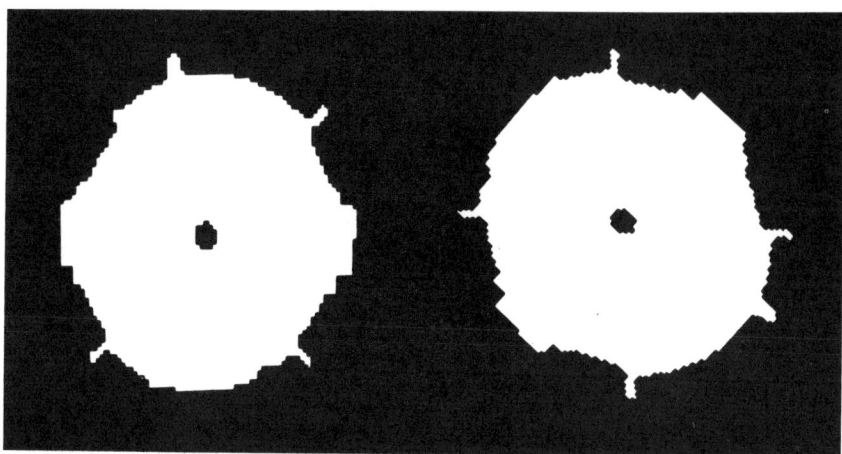

Figure 5-1. Example of complex parts which demonstrate the advantage of machine vision. It is not quickly evident to the human eye that one part has a tab missing. *(Source: Control Automation, Inc.)*

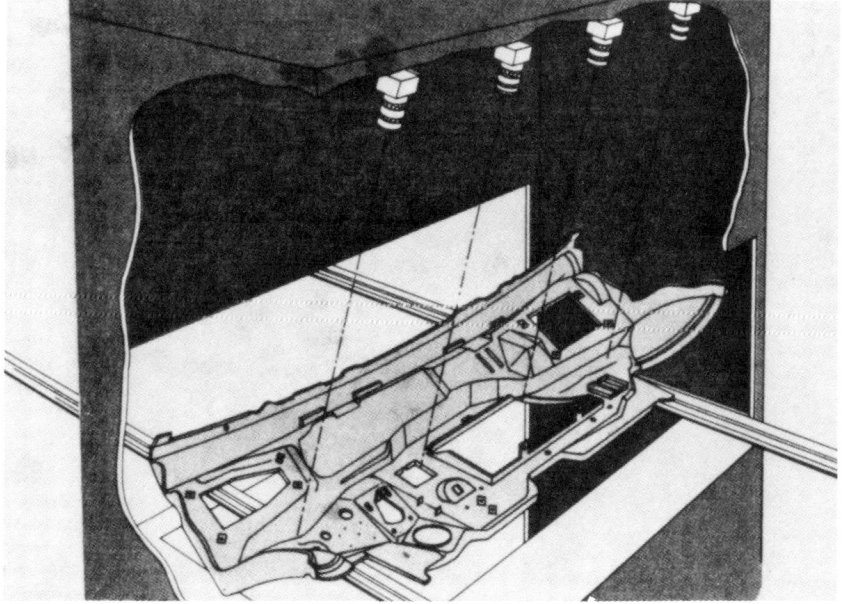

Figure 5-2. Application of Opto-Sense for inspection of automobile dash panel. *(Source: Cooperweld Robotics)*

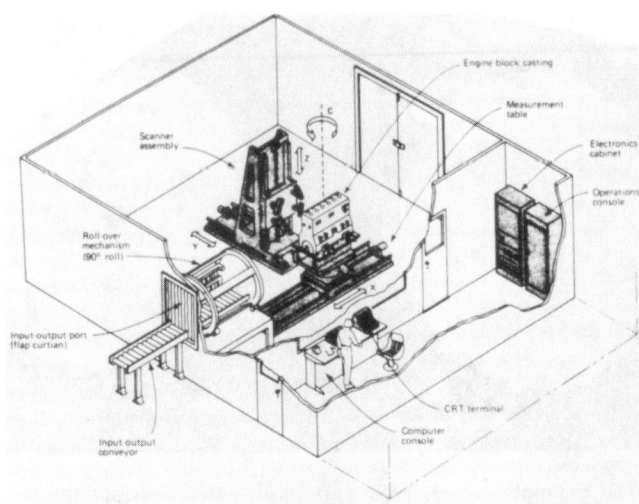

Figure 5-3. Three-dimensional vision inspection of engine blocks at Cummins Engine Co. The ACOMS (Automated Component Optical System) is manufactured by Robitic Vision Systems, Inc., Melville, NY.

INDUSTRIAL APPLICATIONS FOR INSPECTION 65

approximately 40 minutes. The system uses a laser beam light source and a triangulation algorithm to determine distance to the surface at any given point. A set of four fixed sensors scans portions of the surface until the entire surface is scanned. The casting is then rotated 90° and the process continues. The final result is a three-dimensional computer map of the casting, which is then compared with a blueprint pattern stored in a computer to determine errors in dimensional tolerances.

OTHER GAGING APPLICATIONS

The Octek, Inc. Vision System has also been used for visual non-contact measurement. One application is the measurement of a spark plug gap, shown in Figure 6. Manual measurement of the gap is a spark plug is not easily automated with contact guaging equipment. Inspector General's equipment automates feeler guage measurements without precise part fixturing. Non-contact measurements can be made at rates up to 30 parts per second.

Another application[4,5] is a speedometer calibration test, shown in Figure 5-4. A visual inspection system outputs an analog signal to drive the speedometer and tracks the position of the needle using a solid-state camera and image analyzer. The system is capable of tracking the needle over its full range to an accuracy of about 0.3%.

Itran, View Engineering, Object Recognition Systems and several other firms have also delivered similar systems.

VERIFICATION

The application of an Octek Inspector General for inspecting connector contacts is shown in Figure 5-5. Integrated circuit contacts are fabricated using multi-station forming and bending machines.

Defects often go undetected or cause extensive scrap loss prior to detection. Inspector General systems monitor forming or stamping operations and instantly shut down production equipment when a fault is detected. In the figure, the system recognizes the defect as a lack of a hole.

Figure 5-5 also shows the Octek Vision System recording the cutting bar edge dimensions at both ends and at the middle of the surface. Given sufficient processing time, vision systems can make a number of measurements of the images they examine.

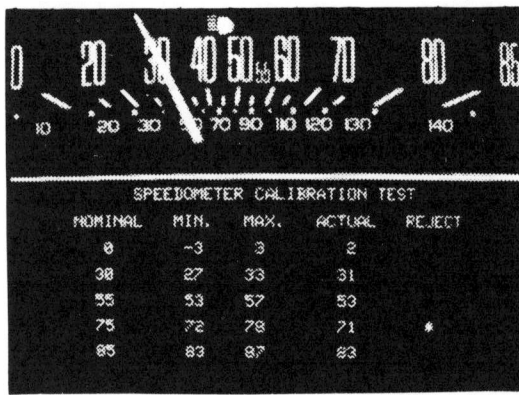 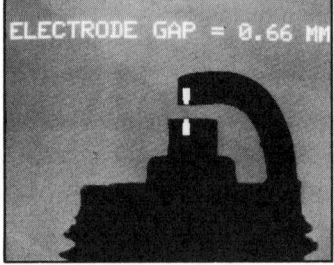

Figure 5-4. Application of the Octek inspector General for non-contact measurement. The measurement of a spark plug gap is shown at the left. A speedometer calibration test is shown on the right.

GRADING FISH AND VEGETABLES[3]

Mitsubishi Electric Corporation has reported the development of pattern recognition software for food processing applications, such as the grading of fish or cucumbers. Typical sorter equipment includes lighting system, sensor (TV camera), picture processor unit, and robotic sorting mechanism linked to the conveyor. As used by one food processor, the equipment can separate up to 14,400 fish per hour into three grades.

In 1980, the Atlanta service center of Western Electric installed a machine to sort telephone parts by color. The system processes 6500 telephone receiver caps an hour with 99.9% accuracy, tripling sorting speed and improving accuracy by 15% over the previous system. The robot's optical head has three photodiodes capped with primary filters to simulate human color reception. A microprocessor analyzes the color signals, then directs the robot to swing its rotating chute into line with one of 12 different color bins. The system is shown in Figure 5-6.

INDUSTRIAL APPLICATIONS FOR INSPECTION 67

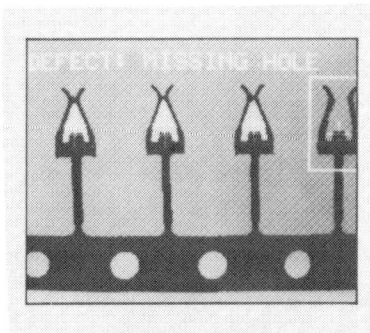

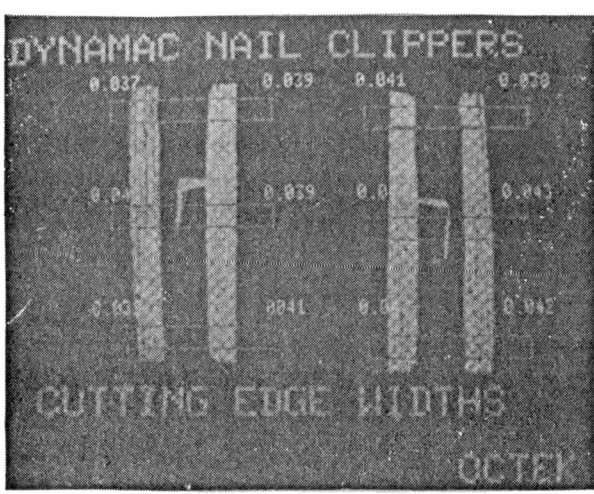

Figure 5-5. Two applications of Octek vision system: (top) inspection of contacts, and (bottom) recording cutting bar edge dimensions.

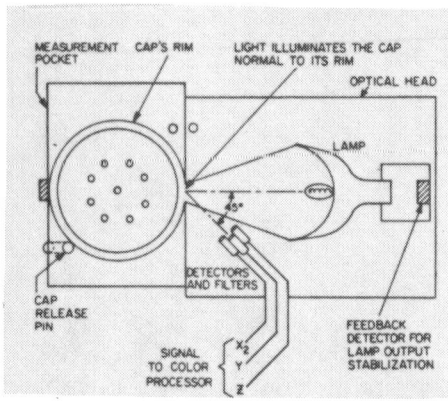

Figure 5-6. Robot vision system at Western Electric for color sorting telephone receiver caps.[6]

CASTING FLAW DETECTION

When a casting is illuminated by light from an oblique angle, it would be expected that a casting defect would cast a shadow which may be detected by a vision system. This technique, however, is found to be only partially successful in detecting larger flaws. Better results are obtained using a laser light, which diffracts at surface irregularities, or by using ultraviolet light.

An example of this class of application is the use of an artificial vision system with a classical magnetic dye penetrant and ultraviolet illumination technique. The forged automotive part shown in Figure 5-7 is magnetized, immersed in the dye, and then placed under ultraviolet illumination and inspected by the Automatix Autovision system to find dye that has seeped into cracks. The high-speed system can simultaneously inspect both sides of this forging at the rate of 60 per minute.[7]

A research team at the School of Engineering, University of Bath, U.K., is currently developing a robotic device capable of removing surface defects from die castings by abrasive machining.[8] They have developed a suitable vision module incorporating a periscopic viewing arrangement attached to a solid-state array camera. Preliminary results suggest that defects can be detected using simple image analysis techniques, provided the surface geometry of the area being examined and the average intensity of the resultant image are taken into account.

GLASS TUBING INSPECTION

Object Recognition Systems, Inc. has recently demonstrated the suitability of the ScanSystem Model 200 to detect bubbles and other flaws in glass tubulation as it is being extruded. Specifically, the application calls for detection of these flaws as the tubulation is extruded at 54 inches per second. Detection is based on dark-field illumination. Only the light scattered by a flaw is observed by the camera.

INSPECTION OF KEYBOARDS

An automatic visual inspection system for quality control of keyboards was described by Joseph Wilder of Object Recognition Sys-

Figure 5-7. High-speed flaw inspection of connecting rods using ultraviolet light and Autovision system.[7]

tems, Inc. at the 1982 IEEE Workshop on Industrial Applications of Machine Vision.[9] The system verifies that each location on a keyboard contains the correct, properly oriented key and that the graphics are not badly distorted. The graphics are inspected on the top and front surface of keys that vary widely in color, contour, surface texture and graphic content. The paper presents economical solutions to a number of problems related to the design of the system's mechanics, optics, illumination, data reduction, pattern location and graphics discrimination. Step-by-step operation procedures for training and running the system are also presented. The illumination arrangement for the inspection system is shown in Figure 5-8. It is reported that the system can inspect a keyboard with 90 keys in less than 45 seconds.

The vision system by Octek, Inc. had also been used for this application. Figure 5-9 shows the vision analysis being applied to a keyboard, detecting a poorly embossed letter.

by Automatix and Object Recognition Systems

INDUSTRIAL APPLICATIONS FOR INSPECTION 71

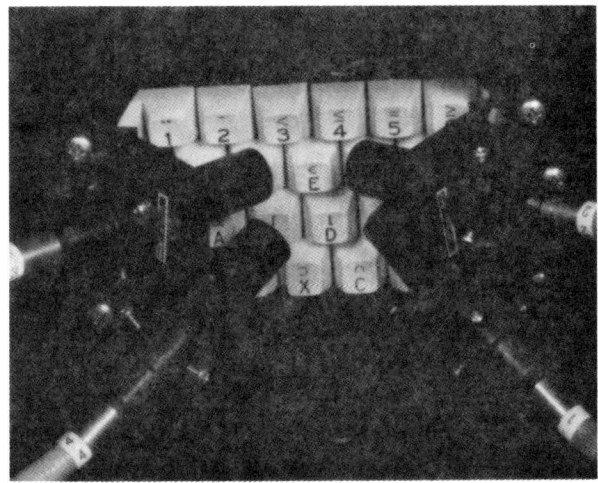

Figure 5-8. The illumination arrangement for Object Recognition System's inspection of keyboards.[8]

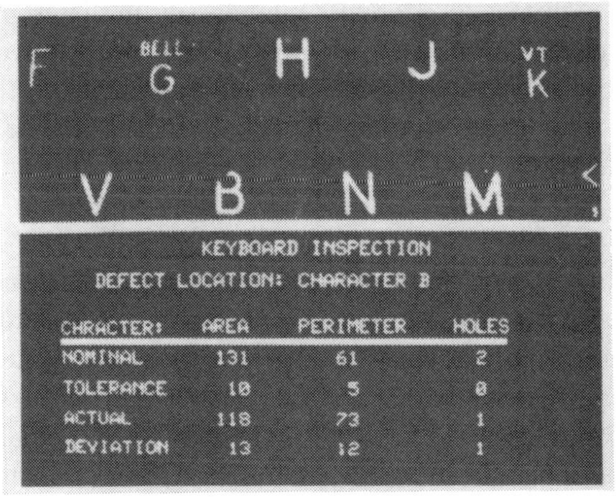

Figure 5-9. Application of Octek vision system to keyboard inspection, showing the detection of a poorly embossed letter "g."

PRINTED CIRCUIT BOARD INSPECTION

The inspection of printed circuit boards is a boring task for human operators and can result in eye strain. Human inspection techniques are also subject to error, especially when the operator is fatigued. For these reasons, machine vision is being applied to the inspection of printed circuit boards (PCB's).

The following summarizes some of the experimental achievements relating to printed circuit boards. The approach has been first to check for the presence and precise location of parts, then to pursue more detailed inspection for part damage and specific identification. The following domain-dependent constraints are used to aid in the inspection.

First, PCBs are composed of relatively few classes of similar components that can be modeled with generic prototypes at several levels of specificity. The inspection algorithm for each class of components can be implemented with a small, modular procedure. That procedure can be invoked for each instance of a member of the class, thereby allowing a large inspection system to be built up from simple and nearly independent modules.

Second, the inherently two-dimensional structure of PCBs implies that the shape of components is unaffected by position and orientation, even though the components may appear in different locations. (This is not always true; components that project above the board on rather "floppy" leads may present different shapes to the camera.) The invariance of shape implies that components can be effectively modeled with simple, static geometric forms that appear unchanged in the image.

Third, PCBs are organized with their components on a rectilinear grid. Because the orientation of the components is known, they need be located only in two dimensions, x and y. Furthermore, these dimensions are usually separable: one can first determine the x location and then the y location, or vice versa. This constraint implies that simple and very efficient one-dimensional signal-processing techniques can be used to analyze intensity profiles from scan lines and scan columns.

Figure 5-10 shows the result of a preliminary investigation at SRI. The system has models that specify the positions and shapes of several parts. Intensity profiles, whose horizontal and vertical coor-

INDUSTRIAL APPLICATIONS FOR INSPECTION 73

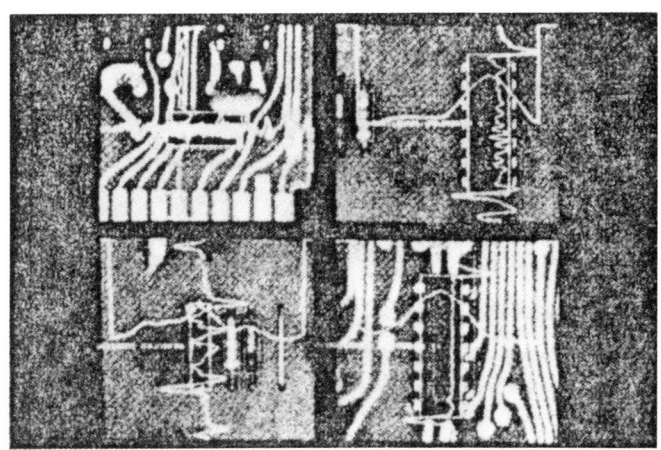

Figure 5-10. Location of printed circuit board parts using intensity profiles.[10]

dinates are selected according to the part shapes, are computed and filtered to enhance certain features from which the precise locations of the parts are determined. The expected and measured locations are compared in the figure.

Machine vision systems applicable to PCB inspection are available from several manufacturers, including Automatix, Control Automation, Object Recognition Systems, and International Robomation/Intelligence.

The Autovision II and Autovision IV systems by Automatix, Inc., Billerica, MA, are applicable to PC board inspection. In small-part electronic applications, as shown in Figure 5-11, a combination of visual inspection and determination of exact location—in this case pin location—is sometimes required to ensure successful insertion. Figure 5-12 shows the verification of relay cans to ensure that the correct part is being inserted, that it is not rotated 180° and, if necessary, to allow an offset to ensure that the pins, as observed by the camera, can be correctly inserted into the printed circuit board.

Control Automation, Inc., Princeton, New Jersey, introduced a high-speed inspection at AUTOFACT 4 (1982). The system is shown in Figure 5-13, along with typical components for inspection. The CAV-1000 Vision System consists of one to four CCD cameras

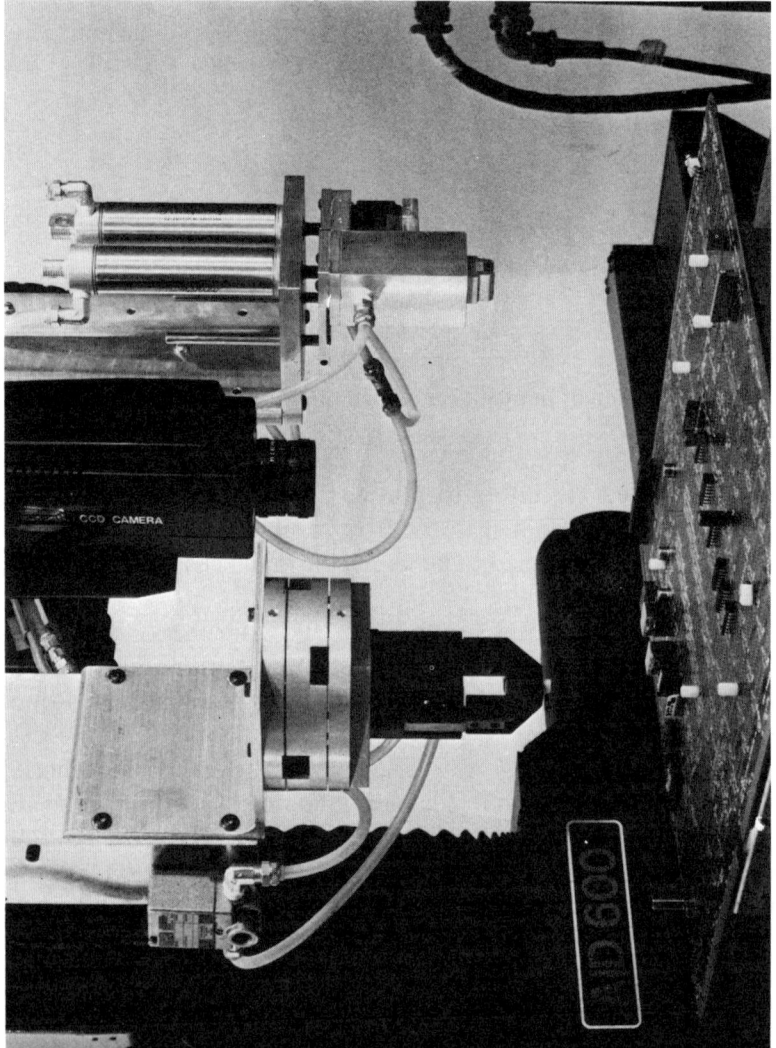

Figure 5-11. Printed circuit board insertion using vision.[7]

INDUSTRIAL APPLICATIONS FOR INSPECTION 75

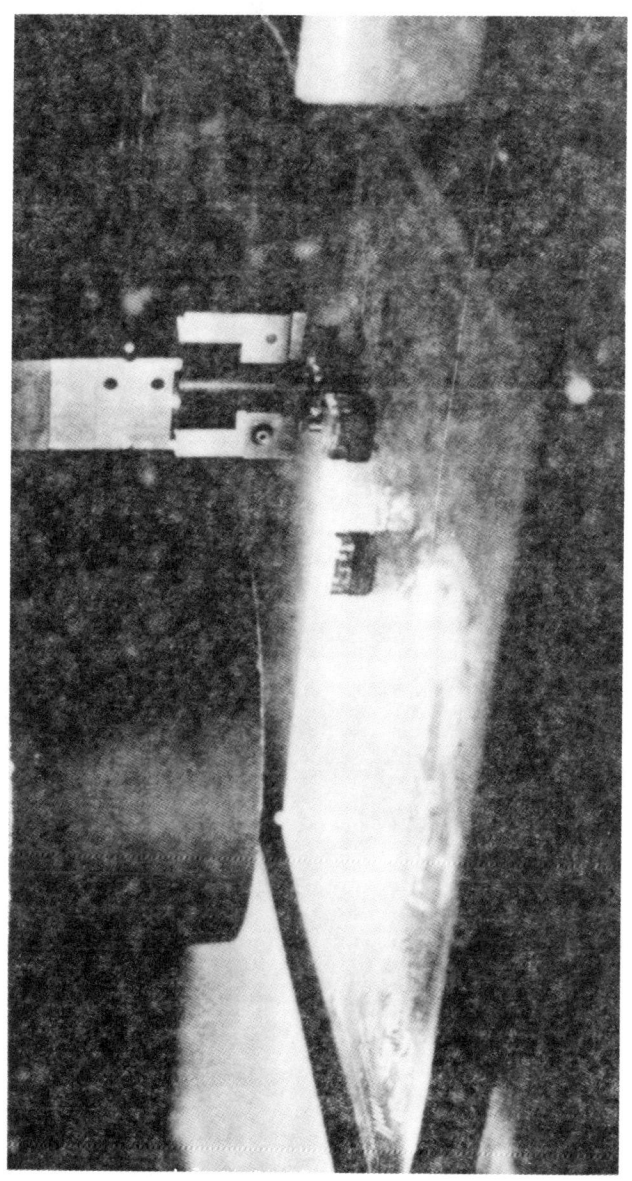

Figure 5-12. Verification of relay cans.

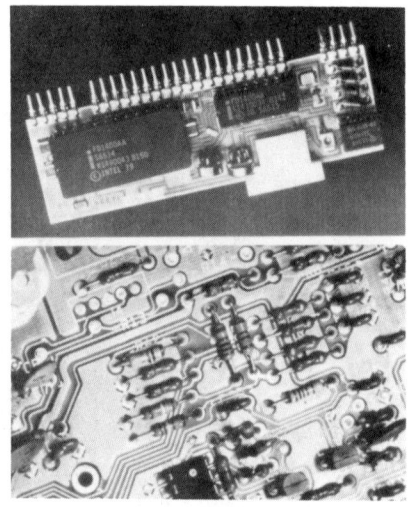

Figure 5-13. The Control Automation V-1000 vision system, shown with some typical components for inspection.

INDUSTRIAL APPLICATIONS FOR INSPECTION 77

with 128 x 128 pixel resolution, a V-1000 Vision Processor and an SC-1000 System Controller. The system has the capability to perform either as a binary or 64-level gray scale system. Speeds of over 80 component inspections per second are attained on such jobs as lead frame inspection for IC assembly and presence/absence decisions for P.C. board inspection. An unlimited number of inspection windows are available with this process. Figure 5-14 shows the system inspecting a printed circuit board for the presence/absence of 137 components. Four windows are used for the inspection,

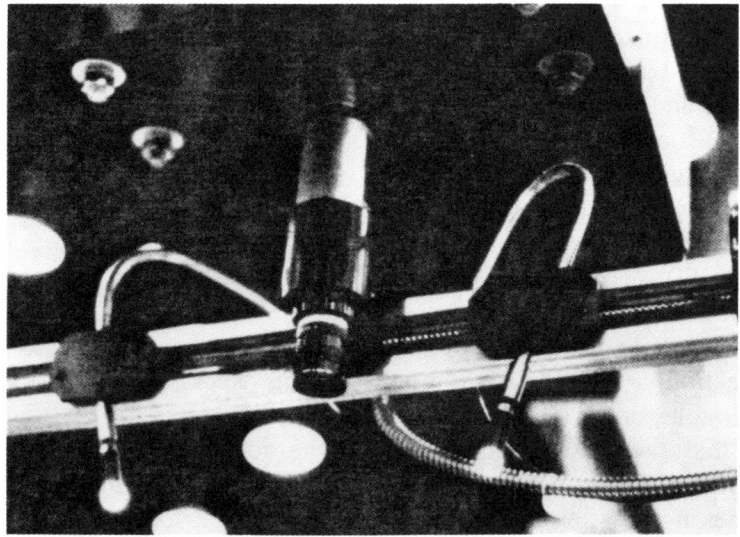

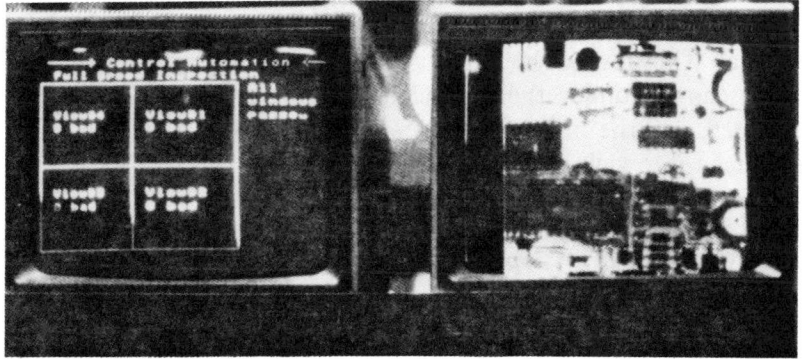

Figure 5-14. The inspection of a printed circuit board for the presence/absence of 137 components. The cycle time including camera movement is 6.1 seconds.

with the camera moving to position for each window view. The cycle time for this inspection, including camera movement, is 6.1 seconds. The system is priced at $19,700 in single lot quantities.

An application of the ORS Multi-Element ScanSystem Model 1000 is the inspection for the absence for components of semi-stuffed or fully stuffed PCBs. The system is manufactured by Object Recognition Systems, Inc., Princeton, N.J. The system, based on proprietary ORS pattern recognition hardware and software, employs an x-y table to segment a scene into discrete geographic elements, each of which is automatically inspected. A micro-computer drives the table to relevant areas based on a library of address information. At each discrete address, the pattern data (collected by the video camera and routinely processed, formatted and digitized by the ORS signal processing electronics) is automatically compared to the stored master "golden" image for that address. "Match/no match" decisions are made and the decision can be displayed and/or printed. The system is designed to address and inspect up to 90 discrete areas within 60 seconds. The ORS Model 1000 is shown in Figure 5-15.

A Vision System manufactured by Octek, Inc., Burlington, Massachusetts, manufactures a vision system applicable to PCB inspection. Figure 5-16 shows the system checking a PCB to verify that a component is inserted, that the component is correctly oriented, and that the inserted component is the correct one. In the figure, the vision system has read the part number on an IC and noted that it does not match the number assigned to that location. The complete PCB inspection system consists of a z axis motor-driven positioning table, solid-state television camera, image analyzer and CRT terminal. The positioning table is used to move a section of the PCB to be scanned under the television camera. Once an area has been inspected by the image analyzer, another section of the PCB is sequenced into view of the camera.

ALPHANUMERIC CHARACTER INSPECTION

The manual reading of labels, lot codes and other alphanumerics is a slow and boring task for human operators. Manual reading is also error prone. For these reasons, machine vision systems are being utilized to automate inspection/verification functions previously requiring a human vision capability. Due to increased efficiency and 100 percent inspection, quality is improved and inventory control is more accurate.

The first industrial vision system for alphanumeric character recognition was DataMan by Cognex Corporation. There are also more of these systems in use today for alphanumeric information than any other model. This system is shown in Figure 5-17. Priced at $30,000, it includes four hardware components: a Vidicon closed-circuit camera, a 14½ x 18 inch RCA monitor and touch-sensitive keyboard and a Digital Equipment Corp. LSI 11/23 processor with a 256K memory and two "frame-grabber" image-capture boards.

A Cognex System is used by IBM for reading the serial numbers on silicon wafers to aid in keeping track of the inventory and quality of semiconductor manufacturing. Because of its advanced image enhancement and optical character recognition algorithms, DataMan can accurately and reliably read laser-etched information such as wafer I.D., vendor I.D., resistivity, etc., from wafers having either matte or polished surfaces (see Figure 5-18).

Four examples of alphanumerics which can be read in industrial environments are shown in Figure 5-19:

- hot-stamped characters on aircraft wires.
- embossed, black-on-black serial numbers on sidewalls of rubber tires.
- expiration dates and commodity codes on pharmaceutical labels.
- stencilled characters on steel billets.

Other companies offering character readers include: Key Image, General Electric and Applied Intelligent Systems.

OTHER APPLICATIONS

A list of additional machine vision applications for inspection was compiled by Gordon Vanderberg and Roger Nagel, while employed by the National Bureau of Standards, in conjunction with Kenneth White of Proctor and Gamble. Information on applications by private companies is often difficult to obtain. Often the very fact that a company is considering an application is proprietary. There are

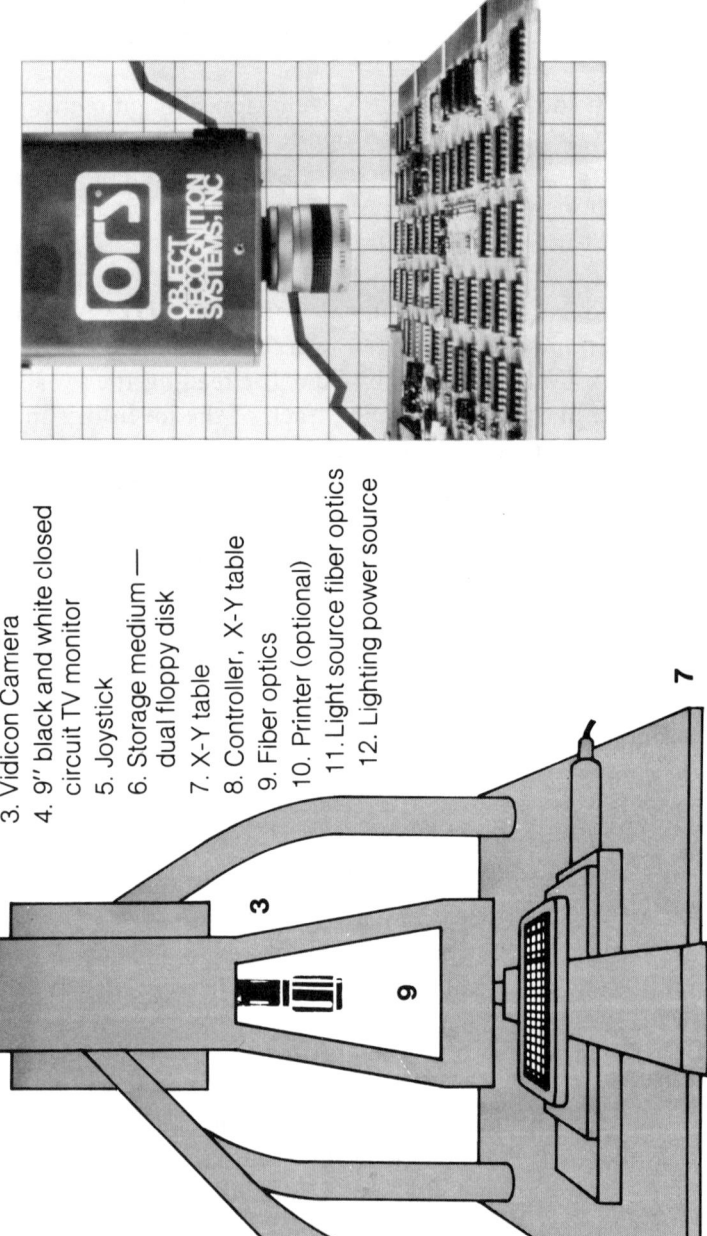

1. ORS Image Processor
2. Interactive Console Display
3. Vidicon Camera
4. 9" black and white closed circuit TV monitor
5. Joystick
6. Storage medium — dual floppy disk
7. X-Y table
8. Controller, X-Y table
9. Fiber optics
10. Printer (optional)
11. Light source fiber optics
12. Lighting power source

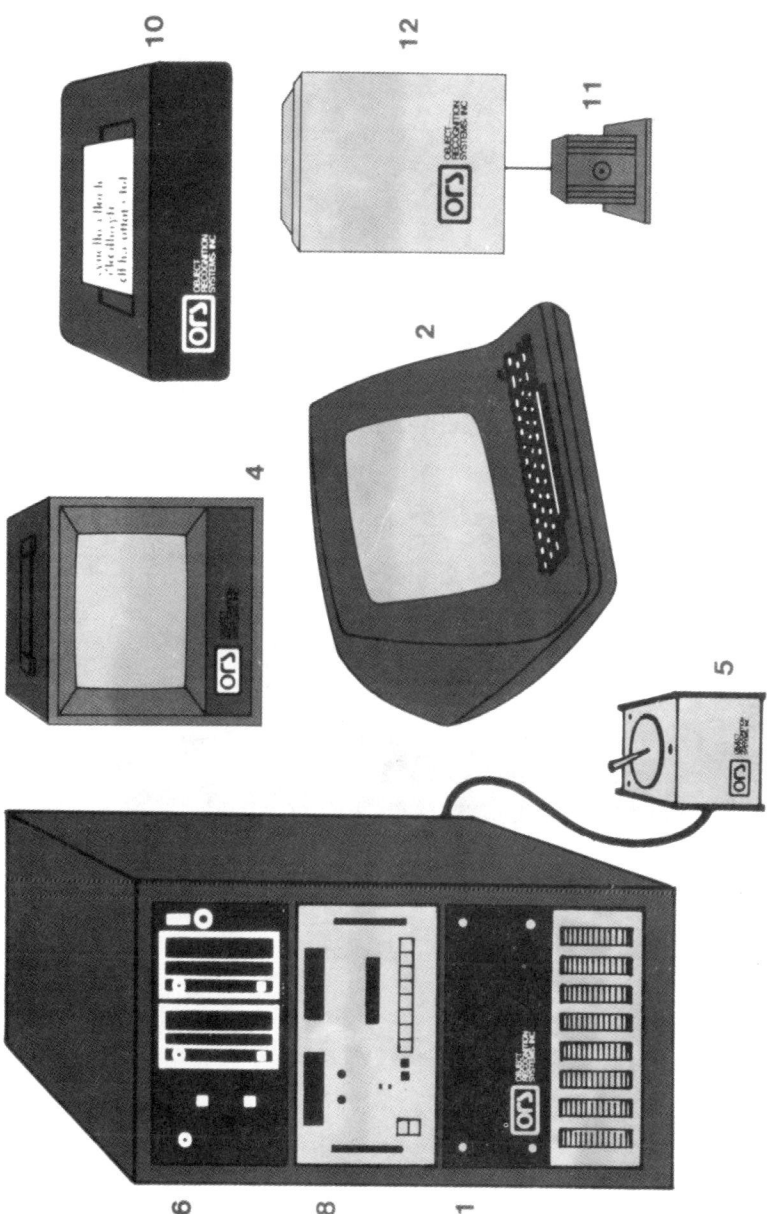

Figure 5-15. The Object Recognition System SCANSCAN Model 1000.

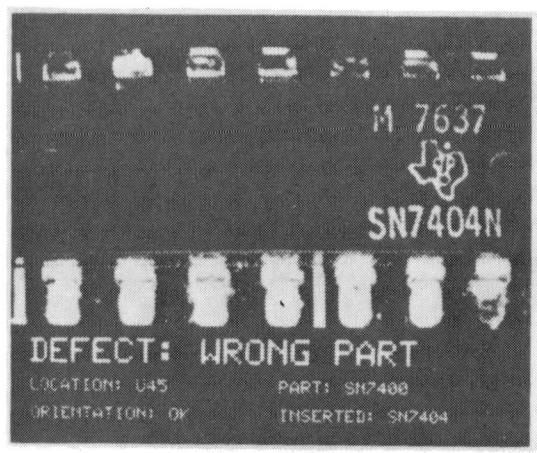

Figure 5-16. The Octek Vision System, shown with an example inspection of printed circuit boards to verify that chips have been inserted into the right sockets.

INDUSTRIAL APPLICATIONS FOR INSPECTION 83

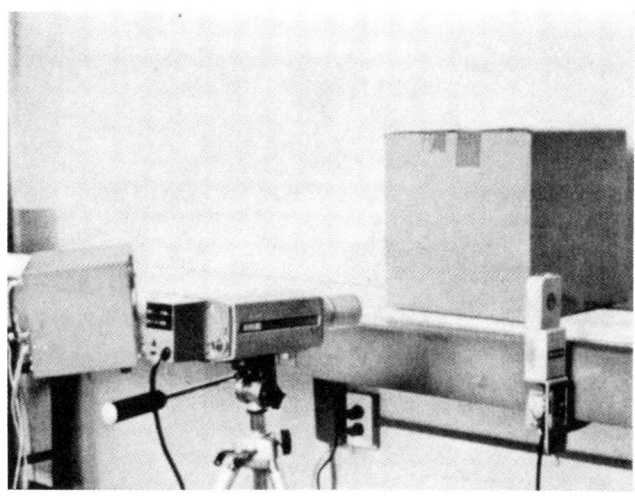

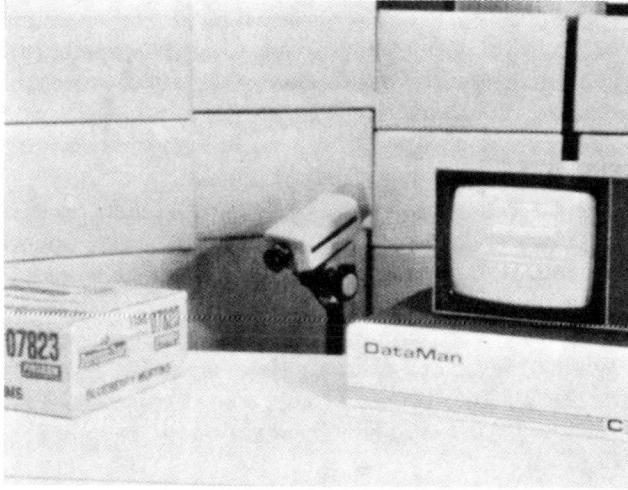

Figure 5-17. DataMan vision system for alphanumeric character recognition by Cognex Corporation.

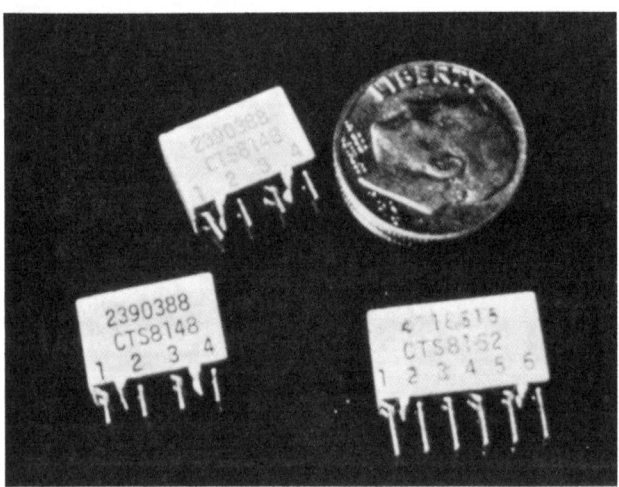

Figure 5-18. Semiconductor manufacturing application: a DataMan reads the serial numbers on silicon wafers to aid in keeping track of inventory and quality at IBM. *(Source: Cognex Corp.)*

INDUSTRIAL APPLICATIONS FOR INSPECTION

no details of the applications available. However, the diversity of the list is, in itself, thought provoking.

1. grasshopper contamination of string beans
2. frozen food quantity count variations
3. dust detection of LSI substrates
4. bottle fatigue detection, to prevent explosions
5. thermometer inspection
6. label inspection on plastic bottles
7. golf ball label inspection
8. IC chip wire bonding
9. measurement of the effectiveness of a window defogger over time
10. automatic debarking, initial timber cutting, sorting of lumber
11. measurement of biscuit height so that packaging equipment does not jam
12. unthreaded nut detection
13. button inspection for the correct number of holes
14. sorting of porcelain seals by shape and configuration
15. automatic closed loop control of extruded gelatin sausage casing
16. hot roll steel web width control
17. particle inspection of pharmaceutical liquids
18. closure integrity inspection for pharmaceuticals
    - is the rubber seal inside the cap provided by the supplier?
    - is the cap tightened?
    - is it cross threaded?
    - is the seal over the cap properly in place?
19. blister pack inspection—is there one pill per blister?
20. china plate inspection for pinholes and bubbles on the plate
21. automated separation of whole almond nut meat from debris, shells, and damaged meat.

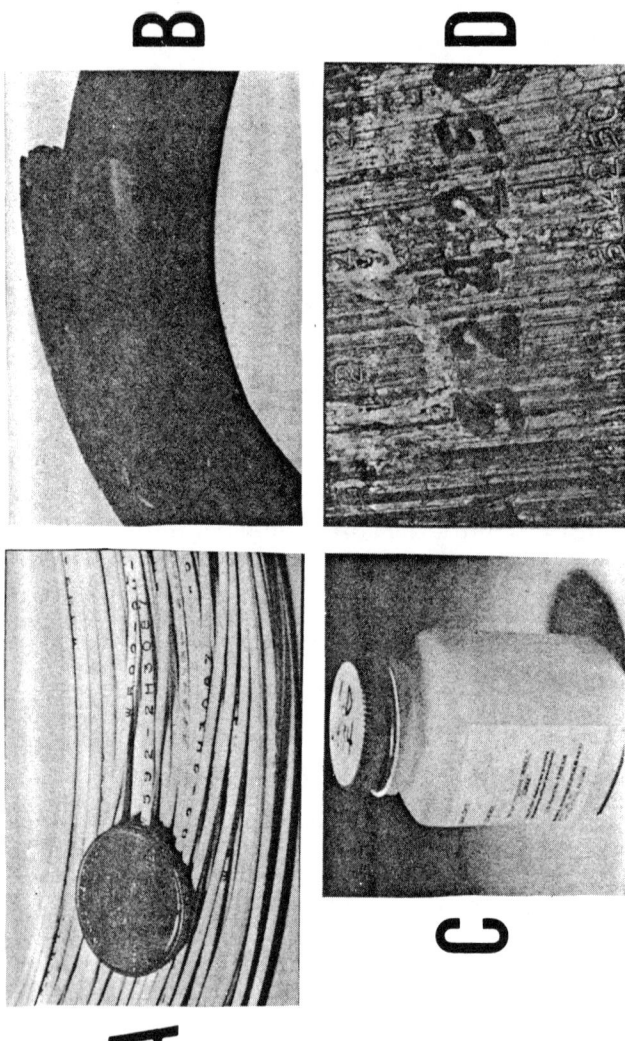

Figure 5-19. Four examples of industrial alphanumeric reading: (A) aircraft wires, (B) rubber tire sidewalls, (C) pharmaceutical labels, and (D) stencilled steel billets. *(Source: Cognex Corp.)*

References

[1] Gevarter, W.B., "An Overview of Computer Vision," National Bureau of Standards Report No. NBSIR 82-2582, September 1982.

[2] Artley, John W., "Automated Visual Inspection Systems Can Boost Quality Control Affordably," *Industrial Engineering*, December 1982, pp 28-32.

[3] Industrail Robots International, November 22, 1982, pg 4.

[4] Trombly, John E., "Recent Applications of Computer Vision In Inspection and Part Sorting," presented at Robot VI Conference, March 2-4, 1982, Detroit, MI.

[5] Trombly, John E., "Adding Machine Vision To Assemble Lines," *Machine Design*, November 11, 1982, pp. 78-81.

[6] Schefter, Jim, "Robots That Think," *Popular Science*, June 1980, pp. 46-47.

[7] Villers, Philippe, "The Role of Vision In Industrial Robot Systems and Inspection," presented at Electro 83, New York, April 1983.

[8] Graham, D. and Choong, Y.C., "Robot Vision In Automated Surfaces Finishing," Proceedings of the 1st International Conference on Robot Vision & Sensory Controls, IFC Conference, 1981, pp. 113-124.

[9] Wilder, Joseph, "Automated Optical Inspection of Keyboards," presented at the IEEE Workshop of Industrial Applications of Machine Vision, May 1982.

[10] Nitzan, David, "Machine Intelligence Research Applied to Industrial Automation," presented at the 9th Conference on Production Research and Technology, Ann Arbor, MI, November 3-5, 1981.

[11] VanderBrug, Gordon, J., and Nagel, Roger N., "Image Pattern Recognition In Industrial Inspection," NBS Report No. PB80-108871.

6

Vision for Industrial Robots

Without sensory feedback, an industrial robot can not intelligently interact with its environment. The most valuable sense that can be provided to a robot, to establish information about the environment and feedback direction control, is vision. At the end of 1982 only a few actual industrial installations of robots using vision were known to exist in the United States. However, those involved in the robotics field had become aware of recent research and product breakthroughs in this area and a rapid increase in the use of robot vision was evident. By 1990, it is estimated that 20% to 30% of all robots will utilize vision.

There are over 100 manufacturers of industrial robots in the United States. The catalogs and product literature of each manufacturer was searched, and the product summaries of the 1983 Robotics Industry Directory were reviewed for robot models with intelligent adaptive response capabilities or the ability to support vision systems. This chapter presents commercially available robots which are either sold with a vision system, or may be interfaced with a vision system of another manufacturer.

- The American Robot Corporation
  354 Hookstown Road
  Clinton, PA 15026
  (412) 262-2085

The Merlin robot is a 60 ips electric driven robot with ±.001" repeatability. The robot can be directed by vision input and has been interfaced with advanced test equipment.

- Anorad
  110 Oser Avenue
  Hauppauge, NY 11788
  (516) 231-1990

The Anorobot features a special multi-tool turret as well as the robotic senses of both vision and touch.

- ASEA, Inc.
  1176 E. Big Beaver
  Troy, MI 48084
  (313) 528-3630

ASEA has over 1200 robots installed worldwide. The introduction of an ASEA vision system was recently announced. This vision system is shown in Figure 6-1 with a model IRb-60 robot.

- Automatix
  1000 Technology Park Drive
  Billerica, MA 01821
  (617) 667-7900

Automatix offers three vision systems for programmable automation: Autovision (Programmable Vision System), Robovision (Programmable Arc Welding System), and Cybervision (Programmable Assembly System). These three systems are shown in Figure 6-2.

Autovision is a programmable image sensing and processing system that inspects, identifies, counts, sorts, positions, and orients parts data. Autovision can use gray scale and up to the 32 cameras to sense defects or to provide vision for robots.

The Robovision programmable Arc Welding System is designed to increase productivity and uniform quality in the weld shop. Optional fixtures and positioners allow manual loading while the robot welds. Over threefold gains in productivity and arc-on times have been achieved.

Cybervision is a programmable system for assembly or material handling. Cartesian and articulated arm robots are integrated with vision and other sensors. Part presentation solutions can be provided as part of complete turnkey and application responsibility. New

VISION FOR INDUSTRIAL ROBOTS 91

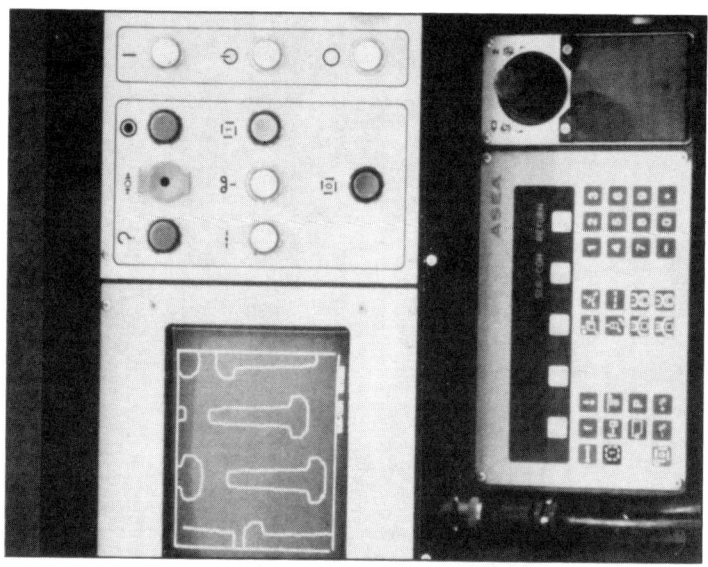

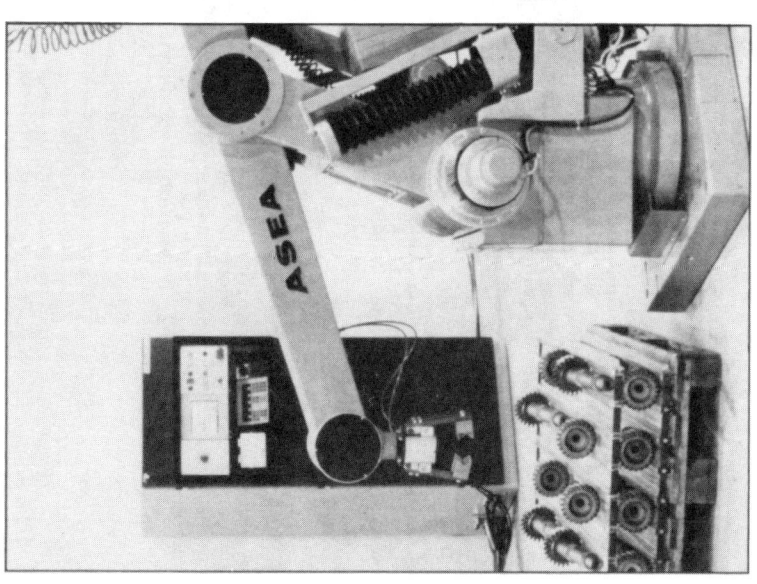

Figure 6-1. The ASEA robot vision system is shown on the left. On the right is the top of the control cabinet showing the monitor, flex-disc unit, and programming unit.

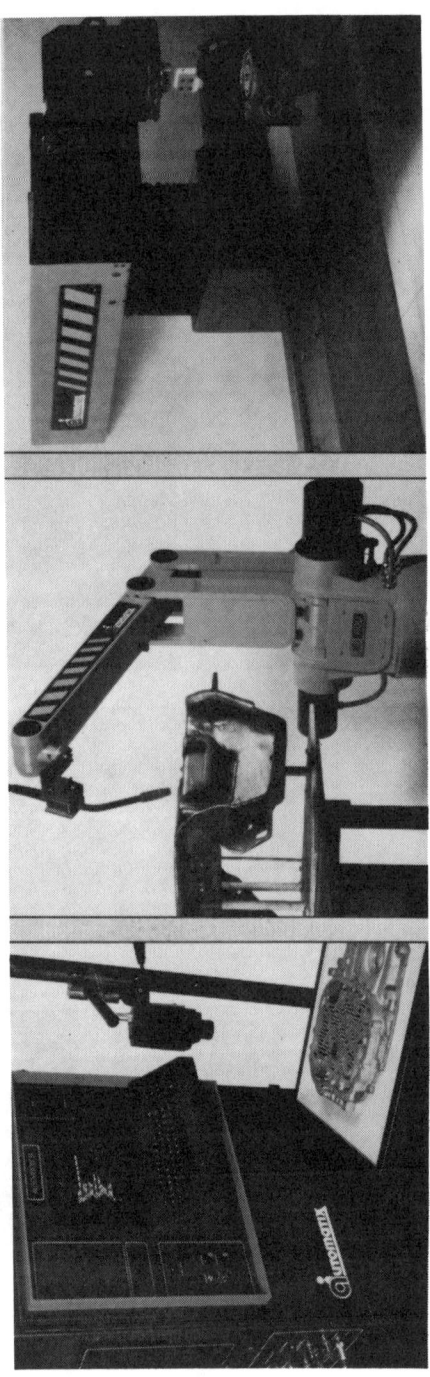

Figure 6-2. Autovision, Robovision, and Cybervision systems by Automatix, Inc.

developments such as CAD/CAM linking and adaptive control will be field upgradable when completed. Cybervision enables the automation of many tasks such as printed circuit boards and mechanical assemblies.

- Cincinnati Milacron
  Industrial Robot Division
  215 S. West Street
  Lebanon, Ohio 45036

Cincinnati Milacron is one of the oldest and largest manufacturers of industrial robots in the United States. Several varieties of their T^3 robot (both electric and hydraulic) are available for applications ranging from welding. The robot can be interfaced with a vision system or weld seam tracking system. Because of the capabilities of the T^3 robot, it has been utilized with several vision systems. The application of a Cincinnati Milacron with the General Motors Consight vision system for sorting castings is shown in Figure 6-3.

- Control Automation, Inc.
  P.O. Box 2304
  Princeton, NY 08540
  (609) 799-6026

The Sembler and Mini Sembler robots by Control Automation can perform tasks such as PCB component insertion under direction of the CAV-1000 vision system (see Figure 6-4). Control Automation also manufactures a stand-alone vision inspection system for parts inspection and a complete vision inspection station for PCB inspection.

- Cybotech Corporation
  P.O. Box 88514
  Indianapolis, IN 46208

At Robots 7, Cybotech's Robotic Vision application demonstrated object recognition coupled with part handling (see Figure 6-5). A camera transmits an image of the object on the conveyor to the object recognition unit, which compares the image to those it has learned. When it gets a match, the object recognition unit signals

Figure 6-3. Cincinnati Milacron T^3 robot. The bottom photograph shows the robot being used in conjunction with a CONSIGHT vision system at General Motors to sort brake drums, connecting rods, and two sizes of castings.

VISION FOR INDUSTRIAL ROBOTS
95

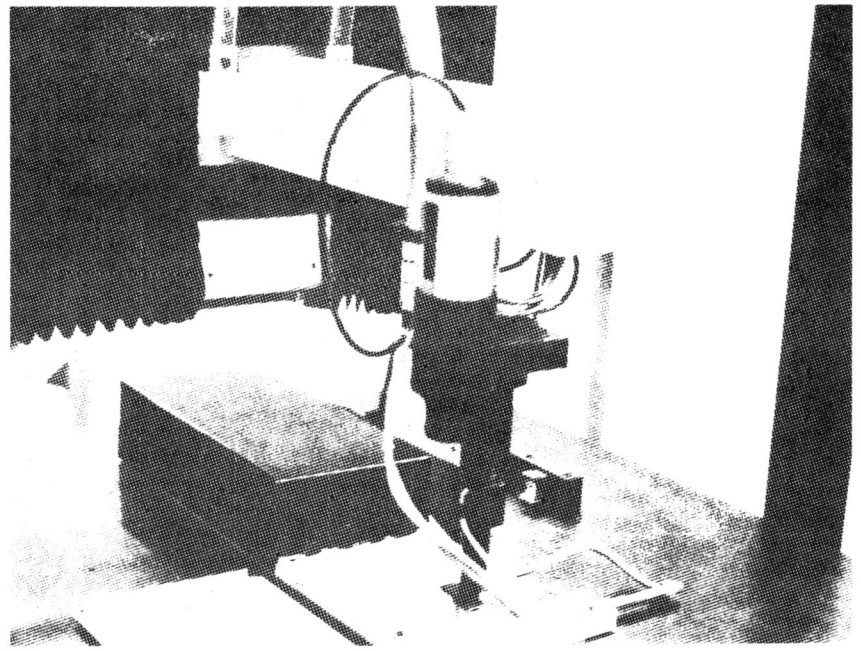

Figure 6-4. The Mini-Sembler robot by Control Automation, Inc.

the RC-6 object controller. The RC-6 directs the robot through the process of approaching the part, adapting the position and orientation of the gripper to suit the position and orientation of the part, then placing the part on the stand.

- Digital Automation Corporation
  65 Walnut Street
  Peabody, MA 01960
  (617) 531-9433

The MACH 1 miniature high-speed robot system can accomodate vision input. The system is designed for semiconductor manufacturing (wire bonding, die bonding, probing, testing, optical measuring and optical inspection).

MACHINE VISION

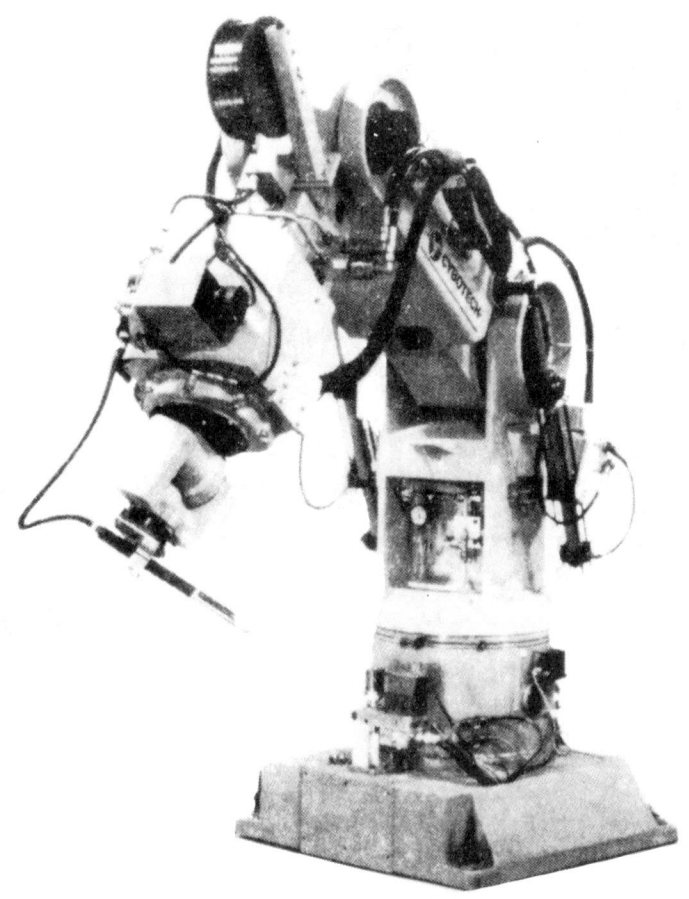

Figure 6-5. Demonstration of the Cybotech Vision System at Robots 7 in Chicago, 1983.

- Everett/Charles Automation Modules
  6101 Cherry Avenue
  Fontana, CA 92335
  (714) 899-2411

Everett/Charles marketed the ERMAC 2500 vision machines in late 1981 and introduced its first robot, ERMAC 26-100 at AUTO-FACT 4. Originally a maker of high-precision machined parts, it has branched into test equipment, specialty electric products and modular automation equipment. The vision system and robot have been interfaced for electronics manufacturing applications. The ERMAC 2500 machine vision system is a micro processor controller with camera that can do real-time inspection of electronics components, stampings and the like. It compares a digitized image with stored standards, and can analyze as many as 1,000 product features up to 60 times per second.

- GCA Corporation
  Industrial Systems Group
  One Energy Center
  Naperville, IL 60566
  (312) 369-2110

GCA Corporation manufactures 12 robot models with vision capabilities for advanced sensory control. At Autofact 4, a Model DKP300H robot interfaced with an Octek vision system, was demonstrated in the assembly of typewriter keycaps.

- General Electric Company
  Automation Systems Divison
  P.O. Box 200, Hwy. 441
  Plymouth, FL 32768

General Electric Company is a manufacturer of both industrial robots and vision systems. GE offers several standard packages which integrate the vision system with a robot, such as WeldVision and BinVision (see Figure 6-6). GE also offers the Optomation vision system for inspection.

- GMF Robotics, Inc.
  5600 New King Road
  Troy, MI 48098
  (313) 641-4242

In 1982, a joint venture between General Motors Corporation and Fanuc, Limited created GMF Robotics. GMF offers 13 robot models which can be directed by a vision system or other advanced sensory device.

Optomation II camera targets part for selection.

Figure 6-6. General Electric's industrial robot directed by a GE Optomation II vision system. This photograph shows the system with the Bin Vision software system for the acquisition of random, ocluded parts.

- Hodges Robotics Int'l Corp.
  3710 North Grand River Ave.
  Lansing, MI 48906
  (517) 323-7427

The Hodges Model MRZ-5 electric robot can support tracking sensors, parts detection, vision systems or proximity detectors.

- Intelledex, Incorporated
  33840 Eastgate Circle
  Corvallis, OR 97333
  (503) 758-4700

Intelledex, Inc. manufactures table-mounted and post-mounted robots (see Figure 6-7) with repeatability capabilities of ±.001" for sophisticated electronics applications. The robots are regularly equipped with integrated vision and an integrated end-effector system for applications which require this flexibility. A stand-alone vision system is also available from Intelledex.

- International Business Machines
  Advanced Manufacturing System
  P.O. Box 1328
  Boca Raton, FL 33432
  (800) 327-0166

The IBM Model 7565 Manufacturing System provides a programmable multifunctional manipulator design for light assembly, fabrication, testing and materials handling. This robotic system is suited for applications where speed, repeatability and product quality are important factors. Tactile and optical sensing features in combination with computer control make the 7565 a very flexible and adaptable system for manufacturing (see Figure 6-8). The IBM 7535 Manufacturing System has been designed for light assembly work as well as other tasks that require speed and high repeatability. It is an electrical drive, microprocessor-controlled robotic system. The 7535 can also be interfaced with a vision system for intelligent and adaptive control

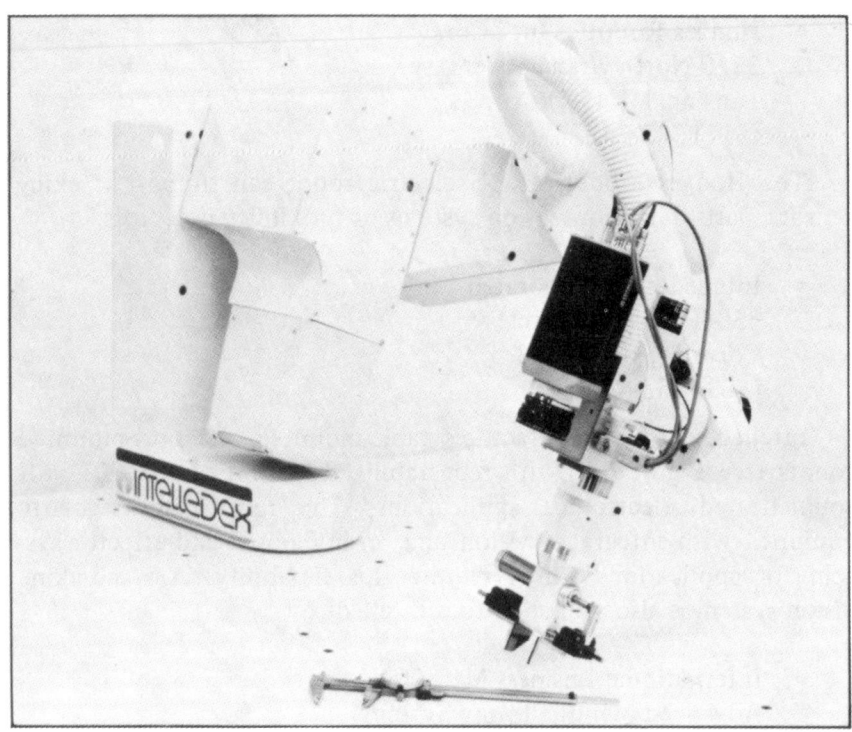

Figure 6-7. The Intelledex Model 605T robot with wrist-mounted CCD camera. The mobil, wrist-mounted camera can often eliminate the need for multiple cameras.

- International Robomation/Intelligence
  2281 Las Palmas Drive
  Carlsbad, CA 92008
  (714) 438-4424

At AUTOFACT 4, International Robomation/Intelligence introduced the Model IRI 256 which works in conjunction with its Model M50 industrial robot (see Figure 6-9). The IRI P256 Vision System is a high-performance gray scale image analysis system for industrial application in robotics, inspection, quality control surveillance, and other areas. The system features a resolution of 256-256 picture elements, the picture element being represented by 256 gray levels.

VISION FOR INDUSTRIAL ROBOTS

The IBM 7565 Manufacturing System can perform a wide range of assembly, fabrication, testing and materials handling tasks... and offers opportunities to set new standards of productivity.

A typical IBM 7565 configuration consists of:
- A manipulator with 6° of freedom (3° to 4° also available) plus optional gripper
- System controller
- Programmable teach pendant
- Hydraulic power unit... plus
- A Manufacturing Language (AML)
- Data communications and remote job entry capabilities

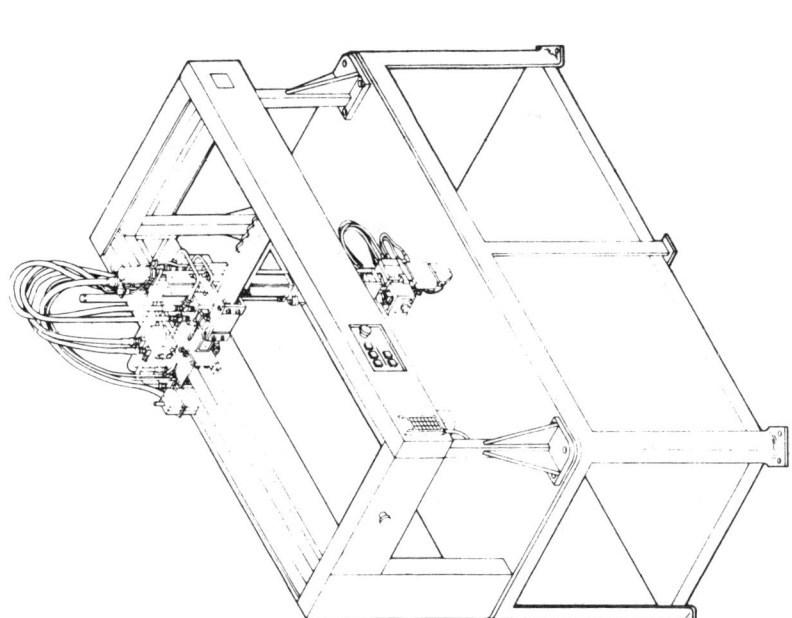

Figure 6-8. The IBM Model 7565 Manufacturing System.

102 MACHINE VISION

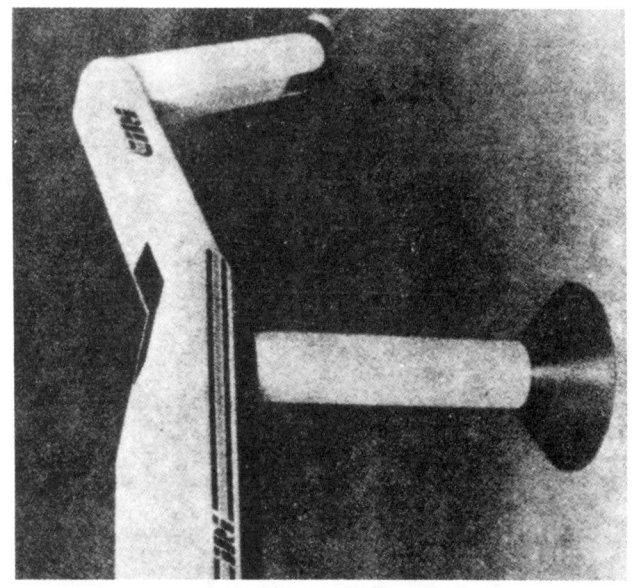

Figure 6-9. IRI Model M50 industrial robot by International Robomation/Intelligence.

- KUKA
  Expert Automation, Inc.
  40675 Heights, MI 48078
  (313) 977-0100

The KUKA Model IR 601/60 industrial robot can support several types of sensors, including vision. The 220-pound-load carrying capacity is unique among robot models which utilize vision.

- MTS Systems Corporation
  P.O. Box 24012
  Minneapolis, MN 55424
  (612) 937-4000

The MTS Model 200A is a general purpose, computer controlled, jointed arm robot. The 200A can accept vision input.

- Prab Robots, Inc.
  5944 E. Kilgore Road
  Kalamazoo, MI 49003
  (616) 349-8761

Prab has over 15 years of experience in manufacturing robots and now have more than 1500 robots installed. Although the firm proposes a "keep it simple" approach to robotics, most Prab models can accept vision input.

- Precision Robots, Inc.
  6 Carmel Circle
  Lexington, MA 02173

The Precision Robot Inc. Model 1000 Robot is capable of up to six degrees of freedom, for such tasks as electronic device insertion/ removal for testing and burn-in, semiconductor wafer handling, PC board testing, and small parts assembly. The model 1000 can be directed by vision system input.

- Seiko Instruments, U.S.A., Inc.
  2990 W. Lomita Boulevard
  Torrance, CA 90505
  (213) 530-8777

Seiko began developing precision mini-robots over 30 years ago to increase internal efficiency and productivity to totally automating the manufacturing and assembly processes of watches. In 1983, Seiko introduced the Model RT-2000 and RT-3000 robots which may be directed by a vision system (see Figure 6-10).

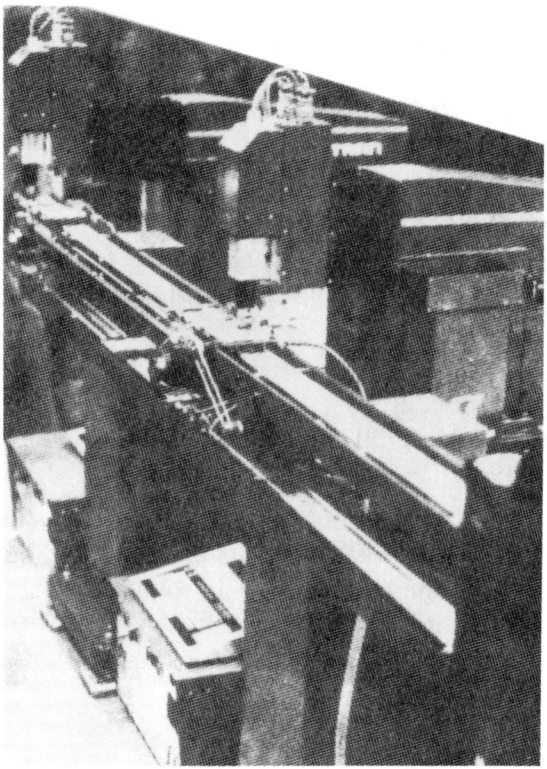

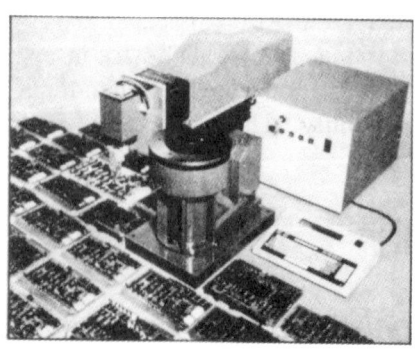

Figure 6-10. Model RT-2000 and RT-3000 robots by Seiko.

- Unimation/Westinghouse
  Shelter Rock Lane
  Danbury, CT 06810
  (203) 744-1800

Unimation, Inc. developed the first industrial robot in 1961, and is currently leading manufacturer in the United States and worldwide. The firm offers Univision I for Unimate PUMA robots.

Univision I consists of a vision processor, graphic display monitor, light pen, camera and Unimation's VAL software and hardware interface. The robot is able to see by recognizing patterns. The vision system locates and categorizes objects within its view based on features of each object's silhouette. For best results, this recognition process requires contrast between the part and the background.

The VAL control system then transforms this information on the orientation of the objects into robot coordinates. This enables the robot to identify the points of an object that it can grasp. The robot then picks up objects, one at a time and moves them as required.

Although the visual system can recognize nine different objects, 12 parts can be identified simultaneously. For each part, 13 distinguishing features can be delineated, including area, perimeter, center of gravity, number of holes and maximum and minimum radii.

Univision I is designed to easily train robots to identify new objects by showing them to the system. The user programming is menu driven using the light pen rather than the keyboard as the primary input. The system is shown in Figure 6-11.

- United Technologies, Steelweld Robotic Systems
  5200 Auto Club Drive
  Dearborn, MI 48126
  (313) 593-9600

United Technologies manufactures and markets in the U.S. the NIKO robots. Seven models have capabilities for sensory feedback and vision control: Articulated Arm Models 25, 50, 150, and 450.

106   MACHINE VISION

Figure 6-11. Univision I system by Unimation/Westinghouse. A Unimate PUMA assembly robot which may be directed by the vision system is shown in the insert.

7
A Summary of Research at SRI International

This chapter is based on research partially funded by the National Science Foundation under Grant No. GI-38100X, and incorporates information presented by Dr. David Nitran at the NSF Grantee's Conference, 9th Conference on Production Research and Technology, November 3-5, 1981, Ann Arbor, Michigan.

It is essential for both economic and social reasons to advance automation in batch manufacturing of discrete products. Our nation is faced with the need to combat inflation and to compete effectively in world markets. We should also make it our objective to improve the working life of the labor force and raise the standard of living of the population as a whole. To fulfill these requirements, we must develop a new technology called "programmable industrial automation." Its salient characteristics are flexibility and adaptability, i.e., the capability of a machine system to perform a variety of tasks under variable conditions, and the ease with which such a system can be trained by factory personnel to perform these tasks efficiently.

In addition to NC machine tools, programmable automation systems existing today include industrial robots that are flexible and easy to train for jobs done under fixed conditions. However, these robots are not adaptable. They are unable to replace human workers who use their muscles, senses, and brains to perform material-handling, inspection, and assembly tasks under variable, often unpredictable conditions. To overcome this limitation, we anticipate the increasing development of intelligent, computer-controlled robots consisting of arms with versatile end-effectors, sensors, and related operational components.

The main objectives of the NSF-sponsored program at SRI International, Industrial Automation Department, are to explore and develop general-purpose and cost-effective techniques and hardware/software modules for computer control of modular systems of manipulators, sensors, inspection, and assembly tasks, and to transfer this technology to industry.

The method of approach to achieve these objectives is as follows:

1. Development of basic techniques and hardware/software modular subsystems for manipulator path control, industrial vision, sensor-controlled manipulation, accommodation, training aids, and distributed microcomputer communications.

2. Integrating these subsystems into a programmable system and experimentally demonstrating the training and execution of material-handling, inspection, and assembly tasks.

3. Affiliation with industrial companies that provide practical, production-related automation problems, assess solutions, and developing technology through mutual visits, consultations, transfer of hardware/software modules, progress report meetings, publications, and films.

TECHNIQUES AND HARDWARE/SOFTWARE MODULAR SUBSYSTEMS

The eleven sections which follow summarize the techniques and modular subsystems of hardware and software (source code in the BLISS-11 language) developed from April 1973 through September 1980.

1. Manipulator Path Control

Analytical techniques and a library of software subroutines were developed to control the trajectory of a manipulator, including (1) transformations between Cartesian and joint coordinates of a Unimate arm; (2) transformations between different stationary or moving coordinate frames; (3) smooth path control for arbitrary trajectories.

2. Analysis of Global Vision Features

Gray-level data acquired from the image sensors are converted into binary form to reduce the amount of image data, thereby increasing the speed and reliability of processing and lowering the cost of equipment. Analytical techniques and library of computer programs were developed for many vision functions to determine the identity, position, and angle of the minimal second moment of objects. These functions include connectivity analysis, feature identification, and training by showing.

3. Vision Module

A hardware/software module for sensing and processing visual images was developed. It can determine the identity, position, and orientation of parts and holes.

This device consists of three major components: a GE Model TN-2200 solid-state TV camera with 128 X 128 elements; a DEC LSI-11/02 microcomputer with a 28K-word memory, storing a library of the vision subroutines previously developed at SRI; an interface preprocessor. The latter component performs the following general-purpose functions in hardware: (1) manual or computer-controlled intensity thresholding that converts the gray-scale image into binary data; (2) a second threshold that, combined with the one above, may be used to generate an intensity histogram of the binary data; (3) run-length coding hardware that converts the binary data into column numbers of the transitions between black and white pixels for each line; (4) two 16K-bit frame buffers, each storing a 128 X 128 binary image, for timing independence between the camera(s) and the microcomputer; (5) up to four TV cameras, each with a strobe lamp trigger signal; (6) a programmable rectangular window for localized processing and display within the image; (7) hardware for accumulating the area and the X and Y first moments of the windowed image, so as to locate its centroid at up to 15 frames per second; (8) miscellaneous functions, such as microcomputer assess to the internal registers and the buffer memories, a phrase-locked loop synchronizing the camera with the ac power, and provision for using a standard oscilloscope as an X-Y monitor.

4. Sensor-Controlled Manipulation

Hardware and software techniques have been developed for using both visual and tactile sensory feedback to control the motion or action of the end-effector of a manipulator. Visual servoing techniques have been explored using an "eye-in-the-hand" (a TV camera mounted on a Unimate end-effector) to reach a target by measuring and correcting the displacement error. Visual servoing with higher contrast and three-dimensional (3D) information was achieved by projecting planes of light on the workpiece, sensing the intersection lines, and processing them. Visual tracking of a free-swinging part was implemented by viewing it with a Vision Module TV camera mounted on a servoed x-y table, and then applying a predictor method to open-loop control of the table at a TV rate.

5. **Analysis of Local Features**

Techniques for automatically extracting local features (e.g., holes and corners) were developed to determine the orientation and symmetries of objects, as well as to recognize and locate partially viewed objects. A matching method, called the Local-Feature Method, has been developed to automatically select the most distinctive local features during system training, thus increasing the speed and reliability of object recognition and location during run time.

6. **Structured Light**

Planar light was applied to inspection of three-dimensional (3-D) objects by comparing model lines with the inspected lines of intersection of the light plane and indexed 3-D objects.

7. **Grey Scale**

A method was developed for using gray scale to inspect registered objects for part integrity. The method utilizes two statistical tests comparing model with inspected intensity images—a one-dimensional histogram that is sensitive to lighting variations and a two-dimensional correlation histogram that is sensitive to part dislocation or assembly misregistration. Both tests are sensitive to a missing part.

8. **Dimensional Measurement**

The mean and variance of the parameters of a circle, a blob, and a straight line that are die to visual quantization errors were calculated. These results were verified by computer simulation and laboratory measurements.

9. **Accommodation**

Uniaxial and triaxial pneumatic accommodation devices were designed and built that can exert a controllable force on workpieces. A computer-controlled passive accommodation capability was also incorporated into an x-y table.

10. Training Aids

Experimentation in the use of voice to instruct a computer controlling Unimate actions was conducted utilizing a commercial, spoken-phrase recognition system.

11. Robot Programming Language (RPL)

A user language called RPL was developed to facilitate the writing and debugging of application programs on a PDP-11/40 minicomputer with an RSX-11/M operating system and two RK05 disks, or on an LSI-11/02 minicomputer with an RT-11 operating system and two floppy disks. These programs are used to control material-handling, inspection, and assembly tasks. Application program statements consist of calls to executable library subroutines or to interpretable, user-defined RPL subroutines. The RPL software includes a compiler and an interpreter. The compiler reads the user program text and translates it into interpretable object code. The interpreter calls library subroutines with appropriate arguments according to the instructions in the object code. Single-step and trace modes of operation permit the user to debug his application program easily and efficiently.

EXPERIMENTAL TASK DEMONSTRATIONS

Between April 1973 and September 1980, the research group integrated the above techniques and hardware/software modules and applied them to laboratory demonstrations in the areas of material-handling, visual inspection, assembly and part presentation. These experiments are summarized in the following four sections.

Material Handling

a. Different parts (connecting rods, water pumps, etc.) were placed randomly on a moving conveyor. As each part passed under the camera, the identity, position, and orientation of that part were ascertained by the vision subsystem. The Unimate was then commanded to track the part, pick it up, and transport it to its destination.

b. Using feedback from a force sensor, the Unimate acquired individual water pumps and packed them neatly in a box.

c. Boxes were placed randomly on a moving conveyor, the vision subsystem determined the position and orientation of each box, and the Unimate packed castings into the box regardless of variations in the conveyor speed.

d. Using an end-effector with four electromagnets and a contact sensor to pick up four separate castings from the top of a jumbled pile in a subsystem had ascertained the stable state, position, and orientation of each casting, the Unimate transported it to its destination.

e. Boxes were placed randomly on a moving conveyor and the visual subsystem determined the position and orientation of each box. The Unimate then placed a stencil on each box, sprayed the stencil with ink, and removed the stencil.

Visual Inspection

a. A group of lamp bases was inspected to verify that each base had two electrical-contact grommets at the proper locations.

b. Washing-machine water pumps were inspected by means of binary vision to verify that the handle of each pump was presented at the proper position. These pumps were later inspected for part integrity; the histogram and correlation statistical tests of gray-scale images were used for this purpose.

c. The inspection method utilizing a planar light was applied to inspection of the end bells of electric motors and curved objects for part integrity.

Assembly

a. Force sensing was applied to detect box cover holes and trigger a one-sided Chobert riveting gun to fasten a rivet in each hole.

b. Unimate wrist, the triaxial accommodation device, was used to insert a retaining ring into a slot on a cylindrical shaft.

c. Assembly techniques were developed, using minimum jigging, which were demonstrated by assembling a compressor cover. The assembly station, controlled by a PDP-11/40 minicomputer, included the following equipment: A Unimate robot under LSI-11 microcomputer control; a gripper mounted on a force-controlled accommodator; a programmable x-y table; an Auto-Place manipulator and auxiliary equipment. After calibration the x-y table, guided by the vision subsystem, moved the compressor housing to a fixed position where the Unimate placed the cover on the housing. The Auto-Place next bolted the cover as the Vision Module positioned the x-y table for each bolt insertion and then inspected the result.

Part Presentation

Five methods were developed for programmable part presentation:

a. Isolated parts on a moving conveyor are recognized and located by a Vision Module, then tracked and picked up by a robot.

b. Semioriented parts in a tote-box are recognized and located by means of an "eye in the hand" to extract global and local features, then picked up by the hand.

c. Jumbled steel parts in a bin are "fished for," separated by limited-sequence manipulator with a multielectromagnet end-effector, placed apart on a table, identified and located by a Vision Module, and picked up by a robot.

d. Parts on a vibratory chute or a conveyor are dropped one at a time, shifted, observed, rotated and, if necessary, tumbled and shifted into a fixed state—then picked up by a limited-sequence manipulator.

e. Parts in a vibratory bowl feeder are inspected by a Vision Module for orientation and integrity and, by means of three air jets, are accepted, rejected, or returned to the bowl.

RECENT RESEARCH RESULTS

Progress since September 1980 was classified into five areas, and is summarized in the following sections.

1. Local Feature Focus Method

Extensions were made to the Local Feature Method, a method for recognition and location of partially visible objects that is based on a set of key local features (such as holes and corners):

- The feature selection process now outputs a description of the key features to be used at run time to recognize and locate touching and overlapping two-dimensional parts. These key features are selected automatically from an analytical model of each part.
- A run time system that uses the pattern of key features produced by the automatic selection process to recognize the locate objects was implemented. This system is currently limited to objects with holes.
- A boundary verification technique has been added to the run time system to verify hypothesized objects.
- A run time routine to locate corners in the input images has been completed.
- The automatic analysis is in the process of being extended to take full advantage of boundary features, such as corners.

As a result of these extensions it is now possible to input analytic descriptions of a few two-dimensional objects and have the system automatically select the distinguishing holes and patterns of holes. This description is passed to the run time system, which can then recognize and locate occurrences of the parts quickly and reliably. The system can even determine whether or not a part is upside down if it is not mirror-symmetric.

2. Visual Inspection of Printed-Circuit-Board Parts

Work has continued on the use of gray-scale techniques for part integrity inspection, concentrating on the inspection of assembled printed-circuit boards (PCBs). The approach has been first to check for the presence and precise location of parts, then to pursue more detailed inspection for part damage and specific identification. The following domain-dependent constraints are used to aid in the inspection.

First, PCBs are composed of a relatively few classes of similar components that can be modeled with generic prototypes at several levels of specificity. The inspection algorithm for each class of components can be implemented with a small, modular procedure. That procedure can be invoked for each instance of a member of the class, thereby allowing a large inspection system to be built up from simple and nearly independent modules.

Second, the inherently two-dimensional structure of PCBs implies that the shape of components is unaffected by position and orientation, even though the components may appear in different locations. (This is not always true; components that project above the board on rather "floppy" leads may present different shapes to the camera.) The invariance of shape implies that components can be effectively modeled with simple, static geometric forms that appear unchanged in the image.

Third, PCBs are organized with their components on a rectilinear grid. Because the orientation of the components is known, they need be located only in two dimensions, x and y. Furthermore, these dimensions are usually separable: we can first determine the x location and then the y location, or vice versa. This constraint implies that simple and very efficient one-dimensional signal-processing techniques can be used to analyze intensity profiles from scan lines and scan columns.

3. A Bowl Feeder With An Eye

The technique of using an ordinary bowl feeder combined with a vision capability for the programmable feeding of small parts was improved. Bowl feeders, aided by mechanical wipers, are typically employed to present small parts in a specified stable state and orientation of automatic assembly machines. The mechanical wipers were replaced with the SRI Vision Module and compressed air hoses, and added computer-controlled gates for part sorting. The resulting system is called the "eyebowl." It is entirely programmable in that it can be easily trained to handle an assortment of different parts. Furthermore, the system can simultaneously inspect those parts and "kick them out of circulation" if they appear to be defective.

4. Modular Programmable Assembly Station

Most of the previous conceptual design of a programmable assembly station was implemented. The station consists of several functional modules: a PUMA 600 robot with a gripper, an Auto-Place limited-sequence manipulator with a servoed rotary base, a servoed x-y-theta table, a Vision Module, and a visually controlled bowl feeder (the eyebowl). Each of these modules implements a specific function, thus becoming a building block of an assembly station. Every module is controlled by a module computer and consists of various devices, each controlled by a device computer or processor. The modules communicate with one another and with the station control computer via a 1-MHz Computrol Megalink communication network using a coaxial-cable bus. The modules, designed to be self-checking, have been provided with a custom non-volatile bootstrap memory to ensure their recovery after a computer failure.

5. Assembly Experiments

A few experiments in programmable assembly of electromechanical devices have been performed using the new programmable assembly system under development. These include the partial assembly of a 220-V interlocking plug and experimentation with snap-in chassis components, such as rocker switches and panel lights. Both assemblies relied on multicamera binary vision for guidance. The plug assembly required cooperation between the PUMA robot and an Auto-Place limited sequence arm during a fastening operation (bolting) and part transfer. The bolting operation was performed with the PUMA holding the assembly and moving it under the Auto-Place's tool. Assembly of a semirigid cradle subassembly supporting the platen in an electronic terminal, as well as on the handling of flexible components (wires and cables) for subassemblies in the same terminal are now being worked on.

6. Semiautomated Hierarchical Planning of Assembly Procedures

The issues involved in semiautomatic programming or robot assembly systems from high-level task descriptions are being investigated. A program being used for this work is called SIPE (System for Interactive Planning and Execution), which is being developed by Ann Robinson and David Wilkins of SRI's Artificial Intelligence Center (AIC) under separate funding. SIPE automatically generates a plan for performing a given task. The plan, in the form of a PERT-type diagram that describes a partial ordering of activities, is displayed on an AED color terminal. Written in the INTERLISP language for the DEC PDP-10 or -20 series of computers, SIPE generates the plan by successive refinement of a high-level plan to lower levels of detail. SIPE utilizes a powerful, taxonomic representation of objects, such as tools and sensors, that can be useful in the automatic allocation of plan resources. A person can interactively help SIPE choose among alternative refinements and suggest changes to correct problems at different levels of planning. The problem of just how to describe to the SIPE planning system several basic sensing and manipulation actions, such as locating a workpiece visually and installing it in a subassembly is being investigated. Initially, it is proposed to generate robot control programs with SIPE in a simple programming language such as SRI's RPL. The latter can then be compiled and interpreted to run the robot system.

CURRENT OBJECTIVES

Under continued NSF grants, the SRI International Industrial Automation Department is exploring the following problem areas.

   a. Continue to incorporate the boundary features (concave and convex corners) into the automatic feature selection program.

   b. Explore the application of gray-scale vision algorithms to general inspection of assemblies for part integrity.

   c. Extend the use of controlled illumination by varying its shape, color, and source.

   d. Expand our programmable assembly station by adding a second PUMA module and part presentation modules. We shall also improve the capabilities of the existing PUMA module by adding contact sensors to its end-effector.

   e. Apply the programmable assembly station to assembly operations involving semirigid parts of electro-mechanical products.

   f. Continue to explore the application of SIPE to semiautomated planning of low-level assembly operations.

8
A Summary of Research at The University of Rhode Island

This chapter is based on research partially funded by the National Science Foundation under Grants No. DAR 78-27337 and APR74-13935, and incorporates information presented by Professors J.R. Birk, J.D. Dessimoz and R.B. Kelley at the NSF Grantee's Conference, 9th Conference on Production Research and Technology, November 3-5, 1981, Ann Arbor, Michigan.

For proper operation, most manufacturing machines require oriented workpieces. The feeding of workpieces is commonly done by human operators. The continued reliance on human labor to perform such tasks, however, holds no promise of leading to significant improvements in productivity, especially when the cost of labor is increasing.

Some workpieces can be oriented by mechanical feeders. Sometimes the orientation of workpieces can be preserved between operations. Presently, industrial robots can do the job when the workpieces are preoriented. Soon a new breed of industrial robot with vision may be used where the other approaches are not appropriate.

Program objectives include the following:

1. Develop useful system architectures and experimentally verify robot techniques to load machines with workpieces which are supplied unoriented in containers.

2. Develop methods to find the position and orientation of workpieces.

3. Develop methods to acquire workpieces from bins.

4. Contribute to an understanding of the relationship between vision algorithms and the constraints of computing hardware.

5. Contribute to the development of general hands and the scientific basis of their design.

Program achievements through January 1981 are described briefly here. Definitions were developed for terms relevant to this research, such as "pose" for position and orientation which has six continuous values in the general case. Several arms and hands were built, culminating in the accurate and reliable URI Mark IV arm, which is an excellent research tool. The use of fiducial lights and TV cameras to permit robots to work on machines, which are not rigidly attached to the robot, was investigated. The concept of specifying positions and orientations to a robot program by manually controlling a set of lights in the field of view of TV cameras was developed for simple pick and place operations and other appropriate applications. A theory was developed to automatically determine the existence and direction of axes of rotational symmetry using multiple views of a workpiece. Various system architectures were conceived, resulting in architectures which could transport nearly all workpieces to machines with the proper pose regardless of workpiece pose in the hand following acquisition from a bin.

Several methods were developed to acquire workpieces from bins. One method used a surface adapting vacuum cup gripper and a vision algorithm that found the center of regions which had intensity values above a threshold and intensity gradient magnitude values below a threshold. The vision located smooth surfaces which were more horizontal than vertical. The surface adapting vacuum cup made the computation of surface angle unnecessary. Another method of acquiring parts used a parallel-jaw gripper and a vision algorithm that looked for opposing edges. This method functioned well on the publicized problem of extracting connecting rods from bins.

Various approaches were developed to estimate workpiece pose with six continuous degrees of freedom. Initial work was done using global binary image features. Images were collected for various orientations of the piece and the image features of a workpiece with unknown orientation were compared to the features of the reference set. The same feature comparision method was also tested using local binary features, such as holes and corners, and using gray scale moments. Another tested method of pose estimation was based on three dimensional models of the position of local features on workpieces. Local features were located in space by using two images of a workpiece and triangulation techniques. Local features were found by computing the gradient direction histogram for an aperture as it was scanned over the images. Sharp corners exhibit two peaks in local gradient direction histograms and small holes have many different entries in these histograms. When the set of feature points in three-dimensional space could be uniquely matched to the feature points model, then the pose could be computed. The mathematics has been developed to compute the constraint equations on workpiece pose variables from surface measurements, such as the three-dimensional coordinates of surface points, feature points, straight edges, and normal direction vectors. A gage was built to accurately measure the pose of workpieces which have a right trihedral corner. It uses 3 pairs of LVDTs mounted at right angles.

An experimental robot system was tested which was designed to acquire unoriented workpieces from a bin, analyze workpiece orientation in the robot hand, and then orient the workpiece on a goalsite. The software had an instruction phase which permitted the system to function with different workpieces, such as a conduit junction box and a gutter end plate. However this complete system could only handle a limited number of different workpieces, primarily because it was based on a vacuum cup gripper and the method of pose estimation was not very general. Nevertheless, this was the first robot system which used vision to acquire a class of pieces from a bin, and furthermore it transported pieces to a goal with a unique pose. A key to the success of the system was the design of a surface adapting vacuum cup hand. In addition to the features described above, it had a contact sensor which could be used to control its depth of motion into a bin. Deep bins could be handled because the cup was moved along the ray in space corresponding to the picture element at the computed holdsite. Another gripper was designed with multiple vacuum cups which could adapt to unknown surface angles. This hand could grasp objects with curved surfaces. It also could function if the grasping surface had several holes.

Other program achievements included a technique for accurately modeling TV cameras via data from two calibration planes; a design of a special image processor for computing local gradient direction histograms; a software package for rapidly extracting straight lines in an image; and a study showing that very low resolution images can yield quite accurate estimates of object orientation about the optical axis.

RESEARCH RESULTS SINCE THE JANUARY 1981 REPORT

The research results are based on the desire to have robots handle unoriented workpieces by methods which are applicable to a wide variety of workpieces and which are reliable, easy to program, fast, and inexpensive. The discussion of progress is organized according to the following topics: an instrumented parallel jaw gripper; performance limits for bin picking; matched filtering for bin picking; pose estimation using iterative image matching; pose estimation using 3-D models with straight edges; pose estimation using appearance sampling and gradient direction histograms; surface normal data via radiometry; inversion of the two-plane camera model; a vision analysis tool; and transportation of workpieces using handoff.

An Instrumental Parallel Jaw Gripper

The acquisition of workpieces from bins can be aided by infrared proximity sensors in the fingers. These sensors can tell the arm when to slow up since it is desirable to approach the bin rapidly to minimize cycle time. These sensors can also increase the probability of a successful acquisition by avoiding pieces at the back of the fingers. A hand has been constructed with proximity sensors pointing downward, sideways, and backward on each finger. Also an emitter and receiver face each other between the fingers to indicate the presence of a piece for grasping. The infrared emitters were pulsed to increase the signal to noise ratio. Typically, sensing distances were .5 to .2 in. All the proximity sensors were multiplexed by circuitry within the hand to minimize the number of wires leading from the hand. The hand also features a gripping force of 100 pounds, swiveling fingertips, an incremental encoder, a tachometer, a permanent magnet DC motor, an overload sensor, and a weight of 3.5 pounds.

Performance Limits for Bin Picking

Methods are being developed as part of this research program to use vision to aid the acquisition of workpieces from bins. Given that certain methods have been developed, the question occurs "How helpful is vision?" Assessing the role of vision is complicated by the fact that successful acquisition depends not only on vision, but also on gripper and workpiece. Two performance bounds were defined to assess the role of vision: the percentage success with no vision, i.e., random picking, and the percentage success with human vision watching only the monitor view of the bin controlling cursors. For the experiments conducted on performance bounds, a parallel jaw gripper was used. For small cylinders (60 mm by 15 mm diameter), the percentage success for blind acquisition was 55 as compared to a 90% success rate using a shrinking algorithm for vision. For a larger cylinder (75 mm by 30 mm diameter), these percentages were respectively 25 and 80. For a connecting rod these percentages were 18 and 70, using a collision front algorithm for vision. With the same test conditions, the objects were acquired nearly every time when a human selected a holdsite on the monitor. This observation indicated that the arm and hand could theoretically do quite well and that the camera was properly calibrated during the testing.

This study implies that vision is not necessary unless cycle time is important. If it is important, then vision can make a significant contribution, particularly for objects which are not readily grasped in many places at arbitrary orientations.

Matched Filters for Bin Picking

To guarantee robot versatility, a general method is being sought for bin picking. Past experience in signal processing has lead to the concept of a "matched filter" for detecting a pattern of noise. Essentially, only those frequency bands where the signal to noise ratio is good are monitored for pattern detection. When a workpiece has an arbitrary pose in space, like in a bin, the computation required for direct matched filtering is huge. The computational burden increases with the number of picture elements describing the scene, the number of elements describing the workpiece (matched filter spread point function), and moreover the number of different views corresponding to all workpiece jaw, roll and spin values. In the particular case of bin picking, simplifications can be made in several respects: (1) matching only the part of a workpiece that a given gripper can grasp; (2) using the most relevant eigenvectors of the matched filter space corresponding to all spin values; (3) scene representation resolution as low as the filters permit.

Successful experiments occurred several years ago at URI for bin picking with a vacuum cup. The filter was matching smooth surface areas where the cup could function. A larger window would identify even better holdsites, matching larger workpiece sections, but this was not implemented as a shrinking technique led to excellent results. In the case of a parallel jaw gripper, matched filtering is more time consuming, than for a vacuum cup, since the pattern to detect is not spin independent. Under certain conditions a base of four vectors allow results similar to a set of 100 matched filters, each tuned for a particular spin value.

Future work includes the design of matched filters picking workpieces from inside (e.g., holes). Nonlinear filters may also be required for those situations where local features diluted in the averaging effect of linear filters, for example, when a small hole on a smooth surface prevents the successful use of a vacuum cup.

Pose Estimation Using Iterative Image Matching

After a workpiece has been acquired, it may have an arbitrary pose in the hand. To transport the workpiece to a goalsite, the pose of the workpiece must be determined. An algorithm has been developed which estimates the pose of a workpiece by matching data extracted from a large number of images of an identical workpiece in known poses. The matching task is complicated by the fact that the workpiece may be partially obscured by the hand and that the workpiece image may be arbitrarily rotated and translated with respect to the stored image.

The algorithm is based on a proposed hardware unit that can shift and rotate a gray scale image and transform it to a set of reduced resolution feature images at the rate of 30 images per second. A second hardware unit will do a pixel by pixel comparison of these images to a set of stored, similarly processed images at the same rate. It should be possible to implement these units using techniques very similar to those used in faster scan image processing and display devices now commercially available. These units are currently being simulated in software. The features used are top, bottom, right, and left edges (thresholded gradients), and gray scale. The software, which simulates the proposed hardware units, reduces 128 by 128 pixel images to five 16 by 16 pixel feature images and does the comparison to 50 sets of stored feature images in about 30 seconds. The remaining portions of the algorithm are designed to be executed rapidly by a general purpose computer.

The first step in the matching of an image of a workpiece held in the robot gripper to a stored image is to compute rough initial estimates of the location and direction of the workpiece in the image. This is done by comparing the feature images of the workpiece in the gripper to a set of feature images of the empty gripper in the same position. Since these computations are done on 16 by 16 pixel images, they can be done quite rapidly in software.

The next step is to recompute the feature images using the direction and location estimates. These feature images should then be in rough registration with one set of stored images. The direction and location values are then iteratively adjusted in an attempt to minimize the difference between the workpiece image and one set of stored images. The set of stored feature images whose error measures show the largest changes during the iteration procedure are chosen as the matching set. This criterion is computed continuously during the iteration procedure and the selection may change.

Pose Estimation Using 3-D Models with Straight Edges

A method has been developed for estimating the six continuous degrees of freedom of an object, even when it is partially occluded by fingers. This method, called the "feature points method," has worked for objects with local features such as sharp corners and small holes. Another method of pose estimation is being synthesized which performs the same kind of pose estimation. Hopefully, when the method is developed, it will run in several seconds on a minicomputer. It should work for various objects which do not have detectable local features.

The basic ingredients of the algorithm include the following steps. An object is held before a TV camera by a robot hand. One picture is taken and then, after the workpiece is rotated by a known angle, another picture is taken. For both pictures, the major straight edges are extracted. Stereo correspondence rules are then used to relate the edges in both images. Each edge in an image leads via a camera model to a plane in three dimensional space. The intersection of the two planes from corresponding image edges usually corresponds to the sharp edge of the object. Such edges can be related to a workpiece model consisting of a line for each sharp, straight edge of the workpiece. A unique relationship between computed and model edges corresponds to a unique estimate of the pose of an object.

During the summer of 1981, the software for the various stages of this pose estimation algorithm was being developed. These stages include the extraction of straight edges in images, the correspondence of edges in two images, camera models to accurately generate the three-dimensional planes from edges to images, and the matching of computed and model edges. Performance evaluation of the algorithm will follow testing of each component, assuming that major computational bounds are not encountered.

Pose Estimation Using Appearance Sampling and Gradient Direction Histograms

Appearance sampling is the technique of taking pictures of an object from many different angles. Basically, one can imagine traveling in a high orbit about a workpiece and sampling the appearance for different values of latitude and longitude. It is not practical to store the image, but to store an efficient representation of the content of the pictures is reasonable. For this study, properties of the gradient direction histogram are being stored because they can be extracted in several seconds on a minicomputer and these properties are related to major geometrical structures. A rapid computation of the gradient direction histogram is possible because the gradient magnitude is only computed for pixels with intensities above a threshold and the gradient direction is only computed for pixels with gradient magnitudes above another threshold. Matching an image of an unknown pose with the appearance sampling data base is expedited by extracting peaks and only checking for registration with initial alignments supplied by peaks as opposed to shifting and matching for each histogram bin.

Surface Normal Data via Radiometry

To acquire a part or to estimate its pose, the high resolution data which is easiest to get is intensity information. However, interpreting information is complicated by the fact that reflected light is a function of light source geometry, surface orientation, surface reflectivity, spectral distributions, etc. Hence, other forms of data, such as range data or surface normal data, which depend almost completely on workpiece geometry, merit investigation for acquisition and pose estimation studies. The procedure for estimating surface normals from intensity data has been studied as part of this research program. Basically, if three images are obtained using three different collimated light sources and a stationary camera, and a Lambertian model for reflectivity is assumed, enough constraints exist to uniquely solve for surface normal vectors.

During this study, the major sources of error in accuracy were identified. Problems result from nonuniformity of the three light sources, dynamic range and calibration of the camera, and light source geometry and location. Procedures for managing these problems were developed. The sensitivity equations which relate computed angles with sources of error were derived. Tables were generated to demonstrate the magnitude of the errors predicted by the sensitivity equations. Tests were conducted using various industrial workpieces. Compared to a mathematical model of one workpiece, a nylon sphere, 85% of zenith angle (off optical axis) measurements were within 10 degrees except at pixels near the periphery. By using the sensitivity equations, the essence of the errors in estimated normals for the sphere data could be explained by small errors in the measurement of the angles of the three light sources and by small nonuniformities of the three light sources.

Some of the limitations of computing surface normals by the radiometric technique include specular reflections, shadows, multiple reflections, and transparent or translucent objects. For bin acquisition, problems with light reflected from other workpieces will cause errors in estimated normals. For bin acquisition, the problem of segmenting the set of normals which belong to one piece or one holdsite is also difficult. On the other hand, acquiring workpieces using this technique with vacuum cup grippers on planer surfaces that have printed patterns appears quite promising because with only intensity information from one light source, printed edges are difficult to distinguish from geometrical edges. The results of experiments on a broad range of workpieces, including metallic, nonmetallic, planar, curved, and textured objects indicate the generality of the technique. Various feature based methods have been suggested in the past for estimating the pose of a workpiece. But no general technique for extracting local features on many workpieces at arbitrary viewing angles has been proposed. The role of surface normal data for six-dimensional pose estimation in a robot hand has not been adequately investigated yet. It will probably present advantages for some pieces and disadvantages for others. For example, if the orientation of a print pattern on a planar surface matters, then an estimation of surface normals alone will not be adequate.

Inversion of the Two Plane Camera Model

As part of the effort to accurately move a robot hand along the line of sight corresponding to a single picture element, a camera model based on data from two planes was developed. This model can be used to accurately define a line in space given an image element. For the two plane camera model, the inverse transformation, namely, going from a point in space to an image element, does not have an explicit form. An iterative procedure has been developed to perform the inverse transformation. Experiments indicate that this method is effective in terms of rapid convergence and accuracy.

Vision Display Tool

When research investigators look at a typical image on a raster graphics display monitor, they lack the ability to see the level of detail at which most vision algorithms operate. To aid our investigators with this problem, a program was written to enhance the details of intensity information within a square aperture, which could be moved about an image under joystick control. The program displays the intensity, gradient magnitude or gradient direction information for each picture element in the aperture as a large square patch on the monitor. The patches have the same spatial arrangement as the pixels in the aperture.

Transportation of Workpieces Using Handoff

After a workpiece is extracted from a bin and its pose in a robot hand is estimated, the problem remains of manipulating the piece so that it may be placed into a machine with the proper orientation. One general robot system architecture which addresses this problem contains two arms. If an arm picks up a piece in such an orientation that it cannot place it into a machine, then it hands off to the other arm. If it still cannot place it, it hands it back and so forth. Obviously, optimization at its simplest level includes minimization of the number of handoffs. Optimization in terms of cycle time is more complicated because it involves a dynamic model of the arms and continuous adjustment over the poses at which handoff occurs. For our initial work on this architecture, we have insisted that the pose of a piece after the first handoff be one of a finite number. Specification of these finite poses includes a subset for which the piece can be placed in a machine. Thus the first handoff should bring the piece to one of the goal poses or to a pose which requires the least number of handoffs to get to a goal pose. A handoff graph which completely enumerates the possible handoffs between the finite set of poses in a hand has been used successfully to minimize the number of handoffs. The first handoff is complicated by the fact that the initial pose can have six continuous degrees of freedom in the hand. Thus it is not possible to build a finite graph which includes collision avoidance information. This problem has been handled by modeling the hands to make sure the fingers do not collide. Collisions between the workpiece and a receiving hand can be avoided by having the piece approach the hand along its axis. Since for handoff it is assumed that the initial estimate of the pose is perfect and since pose estimation algorithms are subject to errors, overload sensors are enstalled at the base of the hands to detect problems and to permit error recovery.

Industrial Advisors Workshop

A workshop was held on Aug. 6, 1981 to discuss the results of the work done on this NSF contract and to seek advice on research priorities. This workshop was attended by industrial advisors, advisors' colleagues, and industrial observers.

PROGRAM OBJECTIVES

In the future, work will continue on the main themes of this research program, namely, the acquisition of unoriented workpieces, pose estimation, and the design of hands and system architectures which aid the acquisition, orientation, and transportation of workpieces. Various approaches to acquisition include: algorithms based on gray scale data, such as ellipse or line extraction or edge propagation with model matching; algorithms based on depth data to find opposing pairs or ridges near the center of pieces; forms of illumination which allow shadows to be exploited; various angles of the hand during approach; and control algorithms which use sensors on a parallel jaw hand for small motions prior to grasping.

Various approaches to pose estimation when there are six continuous unknown degrees of freedom and the workpiece is partly obscured include: expanding the domain of applicability of the feature points method by adding information about major straight line segments and large circular structures; expanding the appearance sampling method to cover partially obscured pieces by making comparisons with local patches of reference data, as opposed to matching in some feature space; conducting tests on the surface point method; and integrating the various methods.

Hand design advancements will come by studying structural details and the use of sensors which yield information on the proximity of pieces, contact, forces on fingers, and overload forces. A system architecture with two arms will be studied because it promises the fast feed times which are required for many applications. As part of this, the transfer of pieces with arbitrary, but estimated, poses between two hands will be studied.

9

A Summary of Research at Stanford University

This chapter is based on research partially funded by the National Science Foundation and incorporates information presented by Dr. T.O. Binford at the NSF Grantee's Conference on Production Research and Technology, November 3-5, 1981, Ann Arbor, Michigan.

The program objective was to formulate fundamental problems in building advanced industrial robots and inspection systems, to analyze and implement solutions, to build experimental systems, and to integrate mechanisms in complete systems to demonstrate ways to do more manufacturing operations more cost-effectively than now possible. We build systems which users can use to set up and program jobs easily. Some major limitations of robots are their lack of sensing and perception and their lack of intelligence. As a consequence of their intelligence, robots are difficult to train. Research is directed toward three capabilities: 1. Force sensing and force sensory/motor control; 2. Very High Level Language for robots including geometric reasoning, geometric and mechanical models, planning, and libraries for assembly; 3. Three-dimensional inspection and vision with geometric models and geometric reasoning.

In typical systems, engineering costs dominate. An obstacle to wide use of robots in manufacturing is the custom engineering required for operations robots can do. Also, there are too many operations robots cannot do. Computers, sensors, interfaces, TV cameras, and vision processors are expected to be inexpensive in the near future. This research concentrates on effective use of computers, vision, sensors, and robots in tradeoffs to cut overall system cost, especially cutting engineering costs of using robots, and to make robots more capable. Results of this research apply across a broad range of sensor-controlled programmable production systems.

ACHIEVEMENTS

This research is part of a broad program in robotics and computer vision started by McCarthy in 1965. The project built up a coherent program of software, hardware and experimentation. A hydraulic arm was built in 1968, the Scheinman Stanford robot in 1970. Paul built WAVE in 1971, our second robot programming system. It was the first system to provide several important capabilities of modern robot programming systems: predictive Newtonian dynamics; automatic planning of smooth trajectories; rudimentary force and rouch sensing used in control; and a macro library of assembly operations. This led to the first computer integrated assembly in 1973, the assembly of an automobile water pump from 10 component parts. The assembly system used force and touch sensing extensively along with vision, a power tool, and several fixtures.

The Unimate PUMA industrial robot is based on the MIT version of the Scheinman arm. The VAL programming system for PUMA, based on WAVE, has been programmed by Dr. Bruce Shimano, beginning during his PhD research at SAIL.

Binford, Agin, and Nevatia carried out a program of vision which led to recognition of complex objects which included dolls and toy horses. Binford and Agin pioneered "structured light" together with independent inventions of Shirae and Will. Binford originated the generalized cylinder representation for geometric modeling of complex objects in 1971. Agin built a program which described parts of complex objects in terms of generalized cylinders with circular cross sections. Nevatia built a system which recognized complex objects. The system built descriptions of objects shown to it on a table. This system recognized objects when they were presented in different orientation with articulation and slight obscuration. Nevatia and Binford introduced new concepts for segmenting complex objects into parts and selecting coherent descriptions of complex objects as structures of generalized cylinders. Instead of matching descriptions of objects with models of all previously seen objects, they introduced indexing concepts to select a sub-class of similar objects for detailed matching. The intent was to structure very large visual memories by using coarse structured descriptions based on an attachment hierarchy on the topology of stick figure descriptions.

Buamgart built GEOMED, a geometric modeling system based on a polyhydral representation, using Euler's theorem and a "winged edge" representation. GEOMED was named for its powerful interactive geometric editor.

THE AL SYSTEM

Paul and Bolles programmed piston-crank sub-assembly (documented on film) and clutch sub-assembly of a two-stroke gasoline engine in WAVE. Those assemblies tested the use of sensing for automating setup and alignment of assemblies, and provided experience in programming assemblies. Based on this experience, a new robot programming system, AL, was designed to provide for: real time control of concurrent multiple assembly devices with sensory/motor control, synchronization and cooperation; data types, linear algebra, and attachment for coordinate transforms; more general trajectories.

Users attach coordinate frames to features of parts. AL automatically keeps track of locations of features of a subassembly when the subassembly is moved. It calculated arm positions from positions of objects to be moved. AL provides local coordinate systems to aid in reprogramming changes in assemblies. AL incorporated state of the art concepts to implement real time operations. Without these constructs, a major part of the programs must be in dirty, unstructured code.

AL was completed in 1977. Finkel, Shimano, and Taylor did most of the initial implementation; Foldman and others completed it and added capabilities for measuring forces and exerting forces and programming language constructs beyond original specifications. An AL interpreter was originated as POINTY by Grossman and Taylor and was redesigned and implemented by M. Gini and G. Gini. Majtaba extended the interpreter greatly and made a film showing its operation. Some user interface aids were added including auto-logging of interaction dialogs and a HELP module with a directed graph structure in which each node corresponds to a keyword and a help message. AL has been used by undergraduate and graduate students for course projects including programming the arm to perform cursive handwriting, small motor assembly, and the interfacing of vision to locate positions of objects. A set of programs was written to test new capabilities and to serve as examples of AL programs. Film sequences of these examples were prepared. Goldman and Mujtaba produced extensive documentation, and AL User's Manual. A series of workshops on software for assembly were held to disseminate the AL design, to provide a forum for discussion of robotics issues and applications, and to compare experience with programming complex robot tasks. Assembly of flashlights were programmed interactively in AL, and a recently released film was made of the assembly.

Two Unimate 500 PUMA manipulators and a Machine Intelligence Corporation VS-100 vision module has been integrated into the AL system and used in assemblies. We have obtained a PDP11/60 for servoing manipulators for AL. A Grinnell GMR-270 is now in use, with 4 channels of 8 bits, each 512 X 512 pixels. This system will also serve as a TV camera input.

Assembly of lawn sprinklers was programmed in 1980. This involved the cooperation coordination of a PUMA 500 arm under VAL and a Scheinmann Stanford arm under AL, and the use of force compliance in the insertion of the stem into the body of the sprinkler. When the PUMA arms were integrated into AL, the program was modified so that the assembly was performed completely under AL.

Goldman began implementation of a minicomputer version of AL in OMSI PASCAL for a standalone PDP11. Capabilities for two arm motions have been tested in several tasks. M. Gini, G. Gini, and Mujtaba have implemented extensive capabilities for source language debugging in the AL interpreter. Runtime path calculation of polynomial trajectories was implemented and integrated into the AL runtime system. Mujtaba designed and began implementation of a part of MSM (Motion Sequencing of Manipulators), his extension of MTM for analysis of human operations. The implementation is embedded in the AL interpreter and MSM primitives are reducible into AL statements which can then be executed directly.

Ohwavoriole described a kinematic analysis of assembly using screw theory. The work was applied to the problem of two pegs in two holes.

The system for programmable compliance and force control has been integrated in the AL and used for several assemblies. It provides compliance which can be modified at runtime.

REPRESENTATION AND MODELING

Models of workpiece and workstation can be obtained from a CAD data base, from teaching with the AL interpreter, from vision and force sensing, from symbolic constraints, and from an interactive modeling system. A body of work is emerging in planning systems for transforming geometric descriptions to robot programs. There operations include: 1. Automated grasping: Paul made a module for choosing grasp positions of polyhedra. Overmyer built GRASP, a rule-based system using geometric models in ACRONYM, to choose grasp positions from specifications of task constraints. Rules calculate accessible faces, secure grasp positions, support and maximal stability. See also work in AUTOPASS as IBM. 2. Collision avoidance: Pieper demonstrated on film a system which guided a manipulator through complex obstacles. Subsequent research has been addressed toward cutting computation cost to enable runtime collision avoidance. Widdoes made an improved collision avoidance program for the first three joints of the arm. Related work is reported by Udupa and Lozano-Perez. 3. Translation of assembly constraints into actions: Taylor wrote a geometry system to translate symbolic constraints of the form "part is against fixture" into a system of inequalities and equalities, to deal with errors in part locations. Fine work elsewhere is reported.

Following work on GEOMED, Baumgart and Miyamoto built the SPI system based on generalized cylinders. Bolles used SPI in Verification Vision to provide models of expected scenes. An extensive effort on the ACRONYM system was initiated to use geometric models in inspection and planning actions. Grossman began a line on investigation of the effects of errors and tolerancing in parts manufacturing and assembly. He used his procedure modeling systems stimulate discrete parts tolerancing in order to model the accumulation of errors in parts manufacturing and their effect on the probability of successful assembly.

Mujtaba analyzed automatic assembly of a pencil sharpener using the Stanford Arm and compared motion times with MTM (Methods

Time Measurement) standards for a human and using assembly primitives developed at Draper Lab. Glaser and Liu conducted analyses of assemblies of a carburetor, a distributor, a fuel pump, and a generator. Two observations were made: 1. If a screwdriver is provided, 93% of operations require only one degree of freedom. 2. An appreciable number of operations involve two manipulators, where the second is used as programmable tooling.

SENSORY/MOTOR CONTROL

Scheinman built a wrist force sensor which is small, light and sensitive. These characteristics are important for work with small arms and small parts in assemblies. The wrist sensor uses 8 pair of strain gages whose outputs are transformed by software to 3 force and 3 torque valves. Shimano and Salisbury wrote software for calibration and force calculation. Salisbury and Mujtaba made an automatic self-calibration program for AL. Several finger force sensors have been developed to augment the wrist sensor.

Paul and Shimano provided analysis of force control based on approximation in joint coordinates. Shimano implemented these concepts in a force control package in the AL runtime system. AL has provisions to read the force sensor, to initialize it, and to exert forces: WRIST(A) returns 3 forces and 3 torques in the array A; SET−BASE initializes zero force and torque levels. Data gathering and graphing capabilities for the wrist force sensor were added to provide tools for development of programs using force. We are accumulating a library of force histories for study of forces encountered during assembly and for force teaching.

Salisbury analyzed active compliance and exerting forces; from these results he built a force control system which is fast, general, and programmable. Active compliance is complementary to passive compliance such as RCC. Active compliance is programmable compliance.

Individual RCC devices must be fabricated for each shaft length and compliance. Active compliance can be programmed in action for shafts with different lengths without changing tools and for a wide range of compliances and force levels while in operation. Active compliance has potential for greater dynamic range than passive compliance because sensitivity is not coupled to mechanical deflection but is determined by solid state sensors so that high sensitivity and high stiffness are compatible.

The programmable compliance system is quite general. Three steps are involved: 1. Choose the center of compliance such that the compliance matrix is diagonal. This is always possible. For example, with RCC, the center of compliance is chosen at the tip of the part to be inserted. Specify six stiffness values. We are looking toward generation of the stiffness matrix from geometric models of objects in assemblies. 2. Calculate the compliance in joint coordinates. 3. Calculate torques in joint coordinates.

In using two arms for handling large or heavy objects, in using complex hands or in interacting with complex objects or with linkages in assemblies, we must provide ways to specify and control the motion in a coupled way. Two arms or three or more fingers provide a very large advantage, a factor of 20 to 100, in torque for assembly or materials handling operations. Complex hands allow reorienting objects and provide a means of sensitive force control. Ishida provided a theoretical analysis including feedforward as well as feedback, thus enabling rapid motion. Roderick performed discrete sampled data analysis of the manipulator control system. Several conclusions came from this work. The sampling rate could be changed to minimize sensitivity to changes in inertia and to minimize their dependence on the sampling time interval.

VISION

Bolles designed and implemented a Verification Vision system for inspection and visual control of assembly, interfaced with the AL system. Verification Vision used gray scale techniques which decrease the requirement for special preparation of lighting and background, and it used three dimensional models of objects. Verification Vision used two dimensional image models for planning efficient sequences of operations.

We have worked toward intermediate level and advanced industrial inspection and vision systems beyond current industrial vision systems, thus with wider applicability in manufacturing. Research has concentrated on: 1. greater capabilities including use of gray scale and stereo ranging, segmented descriptions of shape, and three-dimensional modeling; 2. programming for a great variety of complex parts, using visual teaching from CAD models and teaching by showing three dimensions.

We began work on ACRONYM, a vision system for inspection and for picking parts from bins. ACRONYM combines geometric modeling with state-of-the-art problem solving, thus it is the focus of our efforts in inspection, geometric modeling, and in programming robots using CAD models. Users should program ACRONYM by giving object models and simple task specifications. Geometric models are a natural mode of communication for the user and for ACRONYM. ACRONYM is designed to build up a knowledge base and an experience base for assembly and inspection. ACRONYM has a high level language for models and a geometric editor.

ACRONYM has powerful rule-based "reasoning about geometry." It has models of objects, model of visual observations, models of manipulators, and will have models of force sensor observations. ACRONYM relates object model features to observables (defined by capabilities of actual programs). ACRONYM uses especially those observables which are invariant or quasi-invariant, that is those observables which are approximately stable over a range of viewing conditions or which are approximately stable over a class of objects. For example, shafts of screws appear as elongated ribbons from most viewpoints, for most members of the class of screws.

Efficiency is an important consideration. ACRONYM's models are structured with a level of detail from coarse to fine. This structure permits a coarse to fine strategy for efficient description and interpretation. The rule-based reasoning system also permits a set of rules for efficiency.

ACRONYM was implemented sufficiently to test it successfully on real data in aerial photographs of San Francisco airport. It located an L-1011 aircraft using ribbons obtained from a curve finding program of Nevatia and Babu.

The GRASP system was implemented as a particular rule base in ACRONYM. ACRONYM is generic with respect to object class. Programming for a second member of an object class should require only a small amount of effort after programming a first member.

SIMULATOR

Simulation of manipulators and robot programs allows users to debug manipulator programs without endangering equipment or people. Computer graphics has been used in this way to verify tool cutter paths for numerical control. Simulation enables off-line programming of tasks for robots without removing a line from production. Simulation provides the robot user with a "try before you buy" option.

Soroka developed the SIMULATOR, a program which models the robotic set-up in our laboratory. SIMULATOR is intended as a general simulator for mechanical systems and to interface with several modeling languages (ACRONYM, PADL) and several robot programming languages (AL, APT, VAL). Keystrokes simulate teach-box control of the manipulator models. The SIMULATOR detects collisions between graphical objects and beeps when two objects collide. The user can step forward and backwards through manipulator programs to inspect the command sequence.

The SIMULATOR is implemented in ACRONYM on a DEC KL-10 computer in MACLISP. The system will be transported to a VAX.

EXPERIMENTAL SYSTEM

The two Unimate PUMA manipulators have been retrofit to 6 degree of freedom (dof) devices. We have initiated an informal consortium of PUMA users to instrument PUMAs with electric hands with servo capabilities, and with force sensors. Prototypes for both are being fabricated for test, and arrangements are being formulated to make these modules available to the research community. In order to utilize sensing and servo-controlled hands, suitable software may be necessary. A subset of AL may be interesting for this purpose.

Kerr has designed a hand for PUMA. The design was reviewed and approved by a group of experts whose names are not included in order to avoid the appearance of endorsing the design. Fabrication of a hand for the PUMA robots is now in progress. All of the parts should be ready by the end of September, and the initial testing should be completed by the end of October.

The hand consists of independently servoed finger modules which can be mounted on a base plate. This base plate can then be mounted directly on the PUMA or it can be mounted on the force sensing wrist.

Having the fingers as modules will allow the simple construction of two or three finger hands with the fingers mounted at any desired angle. Having independent servos for each finger will allow the hand to conform to the shape or position of an object without having to reposition the arm. Also, with certain hand configuration (e.g., two opposing fingers), it will be possible to reposition a grasped object without having to move the entire arm. This three finger hand configuration has only three degrees of freedom, of course, and as such it is much more restricted than the versatile end effector being done jointly by Stanford and JPL. That latter three finger hand has nine degrees of freedom.

The finger module itself is a parallel sliding finger pad (like the Scheinman type hand) driven by a ball screw. The finger position is transduced by an optical encoder mounted on the end of the ball screw. The finger pads are removable and are interchangable with the Scheinman hand finger pads.

It is projected that the fingers will be able to exert a peak force of 30 lbs. and a continuous force of 20 lbs. A coefficient of friction of 0.30 between the finger and an object would allow the lifting of a 12 lb. object with two finger contact or an 18 lb. object with three finger contact.

Each finger is 4 in. long with a finger pad travel of 2 in. Two fingers mounted side-by-side make a hand 3 in. wide and 2.5 in. from the wrist to the base of the fingers. Each finger weighs 9 oz. and the base plate weighs 3 oz.

Also in progress is the design of a force sensing finger pad. The goal is to get a finger pad which can measure forces in three orthogonal directions with a resolution appreciably finer than the force sensing wrist.

A design of the force-sensing wrists for the PUMAs is complete, based on the earlier design by Scheinman. We are currently fabricating them. The tentative schedule for instrumenting the PUMAs with wrists is: (a) Machining of metal parts, (b) Application of strain sensors, (c) Electronics designed and fabricated and (d) wrists interfaced and mounted on PUMAs.

The wrists are designed to sense forces up to 40 lbs. and moments up to 110 in.-lbs. with a resolution of better than 0.5 ounce.

AL SYSTEM

Goldman has development of a version of AL in PASCAL well underway. Currently we have a version that runs on SAIL without arm motions and a version is running on the 11/45 under RSX 11/m. The runtime systems will soon be ready. Work has started on adding source level debugging features like those in the previous version of AL. A small subset of AL is planned to run on small computer configurations. Meanwhile, there are implementations of AL at Tokyo in LISP, and at Karlsruhe in PASCAL. Arrangements are being investigated for licensing the portable AL, and to provide a mechanism for disseminating, maintaining, and developing AL.

Craig has implemented a significant part of the PASCAL runtime AL code. The run-time code is on the verge of reaching a limited operational state. By mid-September AL will be able to carry out motion statements for the PUMA arms. Addition of code for hands and Stanford arms will follow. This is somewhat ahead of plans. It had been planned to postpone converting the runtime system to PASCAL until the Pascal version of AL was complete.

The addition of Cartesian motion to AL is in the planning stage. Hooks are being left in the PASCAL run-time AL code specifically for the later addition of the Cartesian mode to AL.

Vistnes interfaced the PUMA manipulators directly to AL by way of the microprocessor bus. AL aims for assembly device independence; this was a good test. The AL part of interfacing the PUMAs was not difficult. This required making a solution for the 5 dof PUMA, which is now being updated to a 6 dof solution. An arrangement was made in which Stanford documented a small, restricted portion of VAL related to communication and calibration, in return for access to the source code. A subset of AL was implemented to carry out these tests.

Craig studied optimization of parameters of kinematics for the finger positions and lengths of the three finger hand design. A program was written to optimize parameters of the hand. The choice of performance index was of interest: to maximize the working volume of the hand in manipulating a small object held between the finger tips. The hand design was based on these results.

Craig also participated in design of control algorithms. Control is based on having active force sensing for tendons. The lowest lev control loops are tendon tension control loops . A hierarchical control system was suggested which at the highest level accepts Cartesian commands in position and force, and resolves those commands into tendon tension setpoints. The system allows control of not only the position and force of the object, but also control along the legs of the grasp triangle. The method also does not require inverse Jacobians, using a method related to that of Khatib mentioned below.

DiVincenzo conducted an experiment to assess the speed of assembly, to assess possible delays in AL, and to investigate provisions which might be added to AL to simplify speed-up of assembly. These studies will be continued and some results incorporated in the new implementation of AL. The following conclusions were reached, based on assembly of an electric motor: (1) As expected, program overhead was negligible. (2) Trajectory calculation required about 2% of total time. (3) Time for automated assembly was 38-41 sec, while time for two subjects with 10 minutes practice averaged 28-6 seconds. Thus, automated assembly took 40% longer than manual. (4) The greatest differences were in grasping and sensing operations. Even with two hands, humans grasped much more quickly and more securely than the arms. Even with the constraints which were used, humans used much more sophisticated sensing and searching algorithms. It is in this area that progress is expected soon. Simply allowing opening/closing of the gripper while the arm is in motion would have some benefit. (5) The time involved in nulling is quite significant. For a single motion, nulling ranges from .3 seconds to 5 seconds, depending on the accuracy of the gross motion, thus depending on the speed of the gross motion. (6) The time was decreased from 70 seconds to about 40 seconds in this study.

Khatib has studied and implemented an alternative formulation of Cartesian control of manipulation. The usual formulation operates in joint space by converting Cartesian commands into joint space commands using inverse kinematics. The process is computationally costly, and gives problems near singularities, where several ad hoc prescriptions are used to patch up paths. The alternative is to control directly in Cartesian space by commanding forces in Cartesian space. Control requires computation of joint torques which are obtained by the transpose of the Jacobian. The advantage is that this method does not require computing inverse kinematics, e.g., inverse Jacobians. Thus, it avoids computation at this phase, and avoids singularities. Its disadvantages appear when compensating for non-linearities in dynamics, inertial and Coriolis terms. The corrections are diffucult to calculate in Cartesian space. One approach is table lookup, which is unsatisfactory in that it requires tables for each load. Other possibilities are under consideration.

SIMULATOR

Since the last report, we have used the SIMULATOR to aid in understanding the design and control of a three-fingered hand. ACRONYM is used to modul the hand, and the user is given keystroke control over joint values. The kinematic solution for the hand has been written in MACLISP to provide simulated teach-box control of finger-tip position in Cartesian coordinates. A rudimentary hand control language (HABDEL) was implemented providing the construct (CONFIGURE HAND TO joint-vector WITH DURATION d).

CURRENT RESEARCH

Stanford currently (1986) has a $2 million per year funded research effort in robotics. Current research projects are:

1. Intelligent Systems for Image Understanding (sponsored by DARPA): this project is aimed at artificial intelligence for geometric representation and geometric reasoning applied to computer vision, especially an intelligent stereo vision system.

2. Flexible integrated systems for assembly (AFOSR): development of high-level programming systems for robots, especially planning level, and vision for industrial parts. The research will contribute to development of a successor to Acronym.

3. Computer-integrated assembly systems (NSF): this project is aimed at developing integrated systems for robots including force control and other capabilities for assembly. It was the original source for development of the AL software system.

4. Dexterity in force control with a three-finger hand (DARPA): this project includes construction and utilization of a dextrous, three-finger hand with nine degrees of freedom. The project includes development of coupled force control of three cooperating fingers in task coordinates, i.e., motion and internal forces exerted on an object. The project includes methods for grasping which do not depend on models, but make use of forces and slipping; it also includes model-based recognition by grasping using quadric surface models.

5. Reasoning about sequences and planning physical actions (NASA): the first part of the project dealt with reasoning about sequences and time. The second part deals with space/time reasoning in the physical world in the presence of multiple actors and concurrent actions.

6. Analysis and experimentation in precision assembly (IBM): this project involves analyzing the operations and economics of a particular assembly, developing software to perform these analysis and the economic models to go into the analysis, together with experimental assembly of a precision device using two arms.

10

A Summary of Research at Purdue University

This chapter is based on research partially funded by the National Science Foundation and incorporates information presented by Drs. Richard P. Paul, J.Y.S. Luh and S.Y. Nof at the NSF Grantee's Conference on Production Research and Technology, November 3-5, 1981, Ann Arbor, Michigan.

The purpose of the Advanced Industrial Robot Control system project at Purdue University is to extend the flexibility and usefulness of currently available industrial robots. Purdue has been extending the use of industrial robots into the area of part assembly by enhancing the robot control system, and by the direct integration of force sensing into robot control. Robots do not provide precise position control, the basis for most hard automation, and must substitute the sense of feel, force, and touch in its place. This sensing ability allows robots to perform tasks for which hard automation is currently required and will also enable them to perform many labor intensive tasks, such as product assembly, for which position Controlled automation is fundamentally unsuited.

INITIAL PROGRAM OBJECTIVES

The major accomplishments for the first three years of the grant are summarized in the five sections which follow.

1. **Motion in Joint Coordinates**
The initial theoretical development work and simulation has been performed in the area of motion in joint coordinates. While robot

rasks may be described in terms of Cartesian coordinates or in joint coordinates, robots are more simply moved in joint coordinates. Given two positions close together in space, a coordinated motion in joint coordinates from one position to the next is a differential approximation to a true straight line Cartesian motion. Over large distances coordinated motion in joint coordinates is as predictable as straight line Cartesian motion. The computations necessary to move a robot in joint coordinates are only those necessary to provide for the coordination of the joints. When robots are to work on moving assembly lines, however, the change in relative position between the robot and its work must be taken into account. This may be done by describing the task in Cartesian coordinates of the robot end effector, to which the relative displacement of the line is added. We describe the conveyor in terms of an array of Cartesian positions; the task is described with respect to the conveyor. These two descriptions may be combined to obtain task positions with respect to the manipulator for a discrete number of conveyor positions. These positions may then be transformed into corresponding joint coordinates. For any given tracking task, joint coordinates for two adjacent positions are obtained: one for a position the conveyor has just passed and the other for a position it has yet to reach. Interpolation in joint coordinates, as a function of conveyor motion, between these points then provides for conveyor tracking. For a typical tracking task, an array of 32 transformations is sufficient to provide for tracking to within nominal robot tolerance. We have been developing a language system, known as PAL, which is in three modules: an editor/scanner, a teach module and an execution module. The editor/scanner module will allow a user to create procedures in PAL, read in such procedures from an external file, edit the procedure, check for syntactic errors and compute transformations while parsing the procedures, and write all procedures into an external file. Procedures are stored in an internal form bearing a one-to-one correspondence to the source text. The source text is replaced by a sequence of symbols with values stored in a symbol table. A teach module has been simulated which will be capable of single stepping through a PAL procedure. If positions are undefined, the manipulator will be placed under control of a joy-stick and the operator requested to move the manipulator to the correct spatial position (teaching by doing). If the position is defined, the manipulator will be moved there slowly under computer control. Auxiliary transformations representing such relationships as tools, grip position, and manipulator placement are defined symbolically. The positions defined by teaching are recorded automatically by amending the transform declarations to include initialization values. On teaching, only those positions through which the manipulator must pass need to be taught. Statements in the PAL procedure referring to compliance will modify the joy-stick control to provide the necessary compliance such that those positions requiring compliance (such as a part insertion) can be attained.

2. Minimum Traveling Time

Conventional manipulator control systems are designed in such a way that the manipulator stops at the end of each path segment. For motions made up from a number of path segments this results in an inefficient operation. By eliminating the need to stop at the end of each path segment and by ensuring that the manipulator moves at maximum velocity and acceleration, the traveling time can be reduced. These requirements as well as other physical constraints have been expressed as a set of inequalities, and approximate programming techniques (MAP) applied in order to optimize the motion with respect to time. However, the convergence of the iterative procedure is not guaranteed. To solve this problem, a direct approximate programming algorithm (DAPA) has been developed which is shown to be convergent.

An alternative approach to optimizing manipulator motion relies on dynamically identifying the slowest of the six joints. This joint is driven independently by a PID controller. Coordination is maintained by driving the remaining five joints at the same relative speed as that of the slowest joint. If one of the tracking joints begins to lay behind, then that joint is exchanged with the present slowest joint to become the motion defining joint.

3. Newton-Euler Formulation of Dynamics and Resolved-Acceleration Control for Manipulators

Position control of a manipulator involves the practical problem of solving for the correct input torques, to apply to the joints for a set of specified positions, velocities, and accelerations. To control a manipulator which carries a variable or inknown load and moves along a pre-planned path, it is required to compute these torques accurately and frequently at an adequate sampling frequency no less than 60 Hz in the case of the Standard manipulator. The same computation is also required if some performance index is to be optimized.

A new approach to the problem has been developed. First, the control technique adopts the idea of "inverse problem" and extends the results of "resolved-motion-rate" control. The method deals directly with the position and orientation of the hand. It differs from other methods in that accelerations are specified and that all the feedback control is performed at the hand level. The control algorithm is shown to be asymptotically convergent. Secondly, a PDP11/45 computer is used as part of a controller which computes the input torques at each sampling period for the control system using the Newton-Euler formulation of equations of motion. This formulation is independent of manipulator configuration. It involves the successive transformation of velocities and accelerations from the base of the manipulator out to the gripper, link by link, using the relationships of moving coordinate systems. Forces are then transformed back from the gripper to the base to obtain the joint torques. Using this formulation, the amount of computation increases linearly with the number of links whereas the conventional method based on Lagrangian formulation increases as the fourth power of the number of links.

A program has been written in floating point assembly language which has an average execution time of 11.5 milli-seconds on a PDP11/45 computer for a Stanford manipulator (only 4.5 milli-seconds is for computing the input torques). This allows for a possible sampling frequency of 87 Hz.

4. RTM, A Technique for Analyzing and Specifying Work for Robots

The language PAL, which is a major thrust of this research, requires that the user know precisely how the robot is supposed to perform a given task, step by step and in great detail. In reality, however, the robot task method has first to be planned and analyzed; alternative approaches have to be examined; and finally a proposed task method can be programmed in detail. Obviously, programming the details of several alternatives merely to evaluate and compare them is a tedious, lengthy, and uneconomical approach. Instead, we have developed a higher level, user oriented technique called RTM (Robot Time and Motion), with the following objectives:

a. Systematically specify a work method for a robot in a simple, straightforward manner.
b. Evaluate a specified method per time, number of step, fixture, tolerance, and other requirements, so the alternative work methods can be compared.
c. Utilize the method specification to develop the robot control program using PAL.

In the scope of this research, RTM, the robot work methods analysis and specification technique has three immediate, useful functions.

First, in order to experiment with PAL and implement it on a variety of actual, industrial tasks to demonstrate its capabilities, this technique facilitates the process of adapting and preparing a task method before actually programming it in PAL.

Second, different tasks can be analyzed using RTM and then programmed in PAL in order to test PAL's capabilities, usefulness, and versatility for different task requirements.

Third, in continuation of the previous function, RTM can be applied in identifying shortcomings and discrepancies in PAL which must be corrected.

In summary, the RTM technique can be used in this research as a front-end tool to enhance implementation and testing of PAL with the overall objective of the correct development of PAL.

A similar approach to the one described above has traditionally been applied in the area of human work methods analysis and design. For example, MTM (Methods-Time-Measurement), and other methods of human work analysis rely on predetermined standard elements of human motions. Such methods are widely used in industry for the purpose of studying and designing task methods. Our work indicates that by learning from this well established industrial practice (in using such techniques in regard to human work), the RTM technique, which is based on standard elements of robot work specifically planned with PAL in mind, can be used for the purposes outlined above.

5. Experimentation on Joint Torque Sensing

A good force sensing technique is imperative to increase the robot's adaptabilities, especially in the area of batch assembly. A fast, reliable force servo system based on the proper sensing techniques can increase the speed of robot operation. However the existing force servo systems are differential approximations or are computationally difficult, making the compliance either slow or approximate. To overcome these disadvantages a simple, high gain, wide bandwidth joint torque servo system has been developed which contains no differential approximations and provides a fast response with no computation problems. The resolution of the sensed torques into control information is much simpler than with a wrist sensor. We have tested a single joint manipulator with the joint sensor to verify the experimental results of the theoretical analysis.

Based on this experiment, two joints of an industrial robot have been redesigned and fabricated to include torque sensing capability by means of strain gauges. The resulting control systems reduced the effective frictional torques of the joints from 1072 oz-inches to 33.5 oz-inches. The stability of the closed-loop systems was analyzed by means of the describing function for limit cycles exhibited in the system, which can be removed by an insertion of phased-lead series compensating networks.

RECENT RESEARCH RESULTS

The following four sections summarize research results within the past year.

1. Scheduling of Parallel Computation for a Computer-Controlled Manipulator

Mechanical manipulators became increasingly important for industrial automation in recent years. The job assignment for an industrial manipulator is usually specified in terms of the path travelled by the hand in Cartesian coordinates. For the commercially available manipulators such as PUMA 250 and 500/600 series of Unimation Inc., the motions are preprogrammed and the driving motors at the joints of the manipulator will furnish sufficient torque so that the hand will travel along a path as planned. However, if the load is unknown and/or variable, the motion of each joint may speed up or slow down and the hand is not able to follow the desired path closely. To overcome this problem, one must repeatedly compute the required, input generalized forces which drive all the joints appropriately. Since the manipulator is a highly nonlinear system, this becomes a difficult control problem. In the available literature, various control schemes were presented which include, for example, resolved-rate control, "computed torque" technique, resolved-acceleration control, etc. In most cases, the control scheme involves the computation of the appropriate input generalized forces using the measured data of the displacements and velocities of all the joints, and the values of the accelerations computed from some justifiable formula. Obviously the execution time for the computation of generalized forces partially determines the feasibility of implementing the control schemes since for real-time control, the computation must be performed on-time. Furthermore, for achieving the convergence of the control algorithm, the computation must be repeated very frequently, preferably at the sampling rate of no less than 60 Hz since the resonant frequency of most of the mechanical manipulators are around 10 Hz.

There are a number of ways to compute the input generalized forces, among which the Newton-Euler formulation is most efficient. With this formulation, the computation is feasible in real time for industrial manipulators even with modest computers such as PDP11's. However, if the computation is done on micro-computers, an improvement on the computational efficiency is desirable to achieve the real time control. Since, using Newton-Euler formulation, the amount of computation is linearly proportional to the number of joints of the manipulator, no simplification can be gained without the sacrifice of accuracy. Intuitively, however, the use of parallel processing will speed up the computation. With the advent of low cost micro-computers, the architecture involving multi-CPU processors is becoming attractive.

For the control system under investigation, one CPU is assigned to each joint, or link. Each CPU has its own memory to store its programs and local data. A serial link manipulator is a complicated mechanism and the dynamic behavior of each link affects that of its adjacent links, i.e., one CPU will use data generated by other CPU's of adjacent links. Thus one common memory is connected between every two adjacent CPU's for storing the common data and necessary information for communication. In addition, the dynamic coupling creates precedence relations among the computational subtasks assigned to various CPU's. This becomes an optimization problem of parallel processing under series-parallel precedence constraints. Methods of solving scheduling problems are available in the literature. Their applicability to this problem, however, is restricted. Because of the complexity caused by the precedence relations, a good scheduling algorithm is difficult to find. The conventional "sequencing problems" technique does not apply since the "no passing" requirement is not satisfied. The PERT and "critical path" technique, the "hard-real-time" method, and the dynamic programming approach were developed for scheduling on a single processor. With multi-CPU's, the precedence relations create idle time intervals for some CPU's while they are waiting for other CPU's to complete the predecessor-subtasks. Dhall and Liu, and Adbel-Wahab and Kameda do not deal with the idle time.

To solve the problem, a method of "variable" branch-and-bound will be developed in this report which schudules the computation of tasks by distributing the load in some sequential order among the CPU's under the series-parallel precedence constraints, such that the time for completing the computation is minimum. It differs from the conventional branch-and-bound technique in that the values associated with branches of the solution tree are not fixed due to variable lengths of idle time. In this technique, the precedence restrictions are taken care of by branching, while the omission of idle times is used in estimating the bounds. Essentially the method includes both forward and backward search procedures. During each forward procedure starting at a branching point, a currently feasible solution path (schedule), which may yield a currently upper bound of completion time that includes the idle times, is determined. The searching procedure then backs to the predecessor branching points to discard, if any, those branches whose minimum value of the execution time is larger than that of the currently feasible solution (schedule). The process repeats the two searching procedures alternatively until all the feasible solution paths are exhausted. The final feasible path which yields a currently upper bound is designated as the minimum-time schedule.

ALGORITHM FOR MULTI-CPU SCHEDULING

Essentially the algorithm is a modification of the conventional branch-and-bound technique. Starting from the initial time (or zero time reference), the forward search procedure is applied to construct a feasible schedule. It establishes an upper bound on the execution time of the optimum schedule. In the feasible schedule, each time instant at which some subtasks have just completed serves as a branching point. By applying the backward procedure, it backs to its preceding branching point. Since it is possible that more than one CPU may have completed some subtasks at a same time instant, more than one group of branches (each group associates with one specific CPU) may be enumerated from one branching point. Now, for each of the alternative branches, the minimum value of the execution times correspond to various groups of branches following that alternative branch is estimated by ignoring all the possible idle times. If the estimated value exceeds the currently established upper bound, these branches are discarded and the process backs farther to the next preceding branching point. Otherwise they will be combed through by the forward search procedure. When a new feasible schedule is found and its execution time is no less than the current upper bound, this feasible schedule is discarded. Otherwise it replaces the current candidate for the optimum schedule and its execution time replaces the current upper bound. The process is repeated until all the branches are exhausted. The last candidate becomes the minimum-time schedule.

APPLICATION TO THE STANFORD MANIPULATOR

A FORTRAN program has been written for the algorithm and applied to the Stanford manipulator. Because of its physical configuration a number of origins of the link coordinates coincide. By analyzing the computational equations, one by one, for this manipulator, it is found that the number of arithmetic operations for these subtasks may not be executed at all. Other subtasks deal with loading numerical data only. These subtasks require an insignificant amount of execution time in comparison with those for floating point additions and multiplications. To solve this example-problem on PDP 11's within an acceptable amount of computing time, it is necessary to simplify the problem as follows. First the algorithm is applied to the computation of $\underline{F}_i$ and $\underline{N}_i$ alone. The the same algorithm is applied to the computation of applied forces/torques. Thus the precedence relations between these two groups of computations vanish. Moreover, some of the subtasks may be combined together. For illustration, it is assumed that the execution time for a multiplication is 50 usec and that for an addition is 40 usec. The time for loading data is assumed negligible.

Using the initially determined feasible schedule, the total execution time for computing F_i and N_i, i = 1, 2, . . ., 6, is 7.69 m.sec and that for applied forces/torques is 3.62 m.sec. The minimum-time schedule requires, however, an execution time of 6.46 m.sec for the former, and 3.24 m.sec for the latter. Thus, for this numerical example, the minimum-time schedule reduces the execution time by 15% over the initially determined feasible schedule. There are 308 multiplications and 254 additions in total, and it requires 25.04 m.sec to process them on one CPU. Thus the level of concurrency is 25.04/(6.46+3.24) = 2.64.

2. Dedicated Microprocessor System

The controller's computer system consists of an LSI-11 microcomputer, which is the host computer to six 6503 microprocessor subsystems. Each microprocessor subsystem controls the motion for a particular joint in the manipulator. Its operating program and servo code resides in a 2K-byte 2716 EPROM. Upon initialization the 6503 microprocessor sets up space allocation for parameter values, commands from the host computer, and two of the stack in 128-bytes of RAM located on board. The microprocessor will then wait in a continuous loop until an interrupt has occurred from either the host computer or interval clock timer to execute the appropriate service routine. Since the main purpose of the manipulator is to study advanced industrial robot control schemes, implementation plays a vital role in the verification of such studies.

The problem that arose with the controller in its original state was that it did not lend itself very easily to program modification at the joint level. The only communication the user had with the 6503 microprocessors was through the host computer. This limited the accessability and/or modification of the servo code to just the parameter values located in the 128-bytes of RAM. If new or different servo code is written for the joint's microprocessor, the test procedure would be as follows:

1. Turn off power to the controller.
2. Remove the 6503 microprocessor card from its back-plane slot.
3. Replace the present 2716 EPROM from the dip socket with an updated version of code in another EPROM.
4. Reinstall the 6503 microprocessor card in the back-plane and turn on the controller to reinitialize the system.
5. Command the joint to move under the control of the new code and watch for errors. (Hopefully without catastrophic movements by the joint.)
6. If there were any problems with the code, try to figure out what and where things went wrong.
7. Correct any mistakes found in the code.
8. Erase and reprogram the other EPROM with the corrected code.
9. Return to step 1.

As one can readily see, this procedure is rather time consuming as well as inefficient.

To remedy the situation a self-contained, dedicated 6802 processor system was added to the controller. It was designed and constructed with the main feature being 2K-bytes of 'dual-ported' or multiplexed RAM. The RAM's function is to emulate a 2K-byte 2716 EPROM that contains the operating system and servo code for the 6503 microprocessor subsystem. A 24-wire ribbon cable from the processor board connects the RAM to the EPROM's dip socket on any microprocessor card. To avoid having to remove the microprocessor card from the backplane each time for inserting the cable, the EPROM's dip sockets were wired to zero insertion pressure (ZIP) dip sockets located on top of the microprocessor cards. When the code in the RAM is correct, the ZIP dip socket on the processor board, is for burning (programming) 2716 EPROMs from the RAM. Thus, the cable is no longer needed to the 6503 microprocessor and can be replaced by the burned EPROM.

The layout of the 6802 processor system consists of a 2K-byte 2716 EPROM that contains the 6802 operating code, a multiplexed address and data bus for switching 2K-bytes of RAM between the 3802 and 6503 microprocessors, and an EPROM burning socket. The multiplexers for the address and data buses to the RAM are changed between microprocessors by a 7474 D-type flip flop. The flip flop is triggered by using the preset and clear entries on the chip. These entries are enabled by decoding the upper three address lines from the 6802 processor. The EPROM burning socket is interfaced through an 8255 3-port I/O.

There are two modes for communicating with the 6802 processor, either through a monitor or an editor. The monitor contains the commands to load the RAM, switch the RAM between microprocessors, program EPROMs, and look at memory. If the RAM should need editing, the 6802 processor's operating code has a simple editor for modifying memory locations.

It should be noted that the 6802 can also load 6800 code into the RAM as well. In this way, a new operating system for the 6802 could be developed while still using the existing operating system and burned into another EPROM. The burned EPROM would then replace the 'old' operating system in the 6802's EPROM.

3. Development of Robot Time and Motion (RTM)

RTM has been developed according to the objectives stated above for robot work specification and analysis. The RTM system is constructed around a list of basic elements. It has been implemented and experimented with on two main robot types: the Stanford Arm and Cincinnati Milacron's T^3. For each, a number of work element modeling approaches have been tried.

Element parametric look-up tables based on mean performance time values; regression equations based on experimental laboratory data; velocity control models, which depend on the precise method by which the robot is designed to move; path geometry, which presently requires relatively detailed specification and motion parameters.

In the numerous analyses of realistic robot task performance were found, that the advantage of the RTM method seems to be that most predicted elements are within ±1% deviation from actual performance, and the infrequent outlays tend to cancel each other in a combined task. Specific results for predicting performance time: For the S.A. with the Path Geometry approach deviations were within -2% to $+12\%$; with the table look-up approach—within ±5%. For the T^3: with the velocity control models—within -2% to $+3\%$.

11

Research at the National Bureau of Standards

This chapter was presented at the 10th International Symposium on Industrial Robots, Milan, Italy, March 1980. It was prepared by United States Government employees as part of their official duties and is therefore a work of the U.S. Government and not subject to copyright.

For robots to operate effectively in the partially unconstrained environments of manufacturing, they must be equipped with control systems that have measurement and sensory capabilities. This chapter presents a model for such a system. It consists of parallel control and measurement hierarchies. The control hierarchy decomposes tasks into subtasks, and the measurement hierarchy analyzes data from sensors. At each level the control hierarchy sends expectations to the measurement hierarchy, which returns computed values of the deviation between the observed and expected data. The control hierarchy uses this information to modify its task decomposition strategies so as to generate sensory-interactive goal-directed behavior. The system has been partially implemented on a research robot using a network of microcomputers and a real-time vision system mounted on the robot's wrist.

INTRODUCTION

Clearly, sensory measurement capabilities are necessary for robots to operate effectively in unstructured environments; but more than just measurement data is required. The data must be processed and analyzed and the results introduced into the robot control system in real-time so that the response is goal-directed, reliable, and efficient. This is a problem in which complexity grows exponentially with the number of sensors and with the number of branch points in the control program. The problem of controlling a sensory-interactive robot is similar in many respects to that of controlling any complex system such as an army, a government, a business, or a biological organism. The command and control structure for such systems is invariably a hierarchy wherein goals or tasks selected at the highest level are decomposed into sequences of subtasks which are passed to the next lower level in the hierarchy. This same procedure is repeated at each level until, at the bottom of the hierarchy, there is generated a sequence of primitive actions each of which can be executed with a single operation. Sensory measurements are fed back into this hierarchy at many different levels to alter the task decomposition so as to accomplish the highest level goal in spite of uncertainties or unexpected conditions in the environment.

The advantage of hierarchical control is that complexity at any level in the hierarchy can be held within manageable limits irrespective of the complexity of the entire structure.

Part of the robotics research program at the National Bureau of Standards (NBS) is an investigation of methods for designing and implementing real-time sensory and control functions in a hierarchical structure.[1,2,3,4] Currently, efforts are being directed toward developing engineering procedures for partitioning tasks into subtasks, assigning subtasks to logical modules, designing hardware to implement this logic, and writing software.

HIERARCHICAL TASK DECOMPOSITION

In general, the behavior of a robot results from a time series of primitive actions which generate drive signals to actuators producing forces and movements. Each primitive action is typically specified by an instruction such as GOTO (point) which resides in a memory device. A sequence of instructions (i.e., a program) which executes a complete task can be represented as a sequence of states in a state diagram (5) such as shown in Figure 11-1.

In the absence of sensory input, the sequence of primitive actions is fixed, both in its order and its timing. However, most present day

RESEARCH AT THE NATIONAL BUREAU OF STANDARDS 161

robots allow external interlock signals to synchronize the timing of the robot's program with external machinery.

More sophisticated robots have the capacity to use sensory interlocks to activate conditional branches within the program. This gives rise to the type of state graph illustrated in Figure 11-2. Conditional branching enables a robot to select one of several different programs depending upon sensed conditions in the environment. In some cases, such a robot can be programmed to cope with simple error conditions.

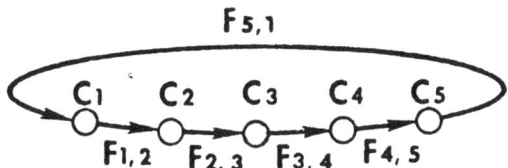

Figure 11-1. A state-space graph representing a simple robot program. Each state Ci corresponds to a single instruction, or primitive action in the robot's program. Fij represents the logical conditions required for the state transition from Ci to Cj. The state transition corresponds to the program counter stepping from one instruction to the next.

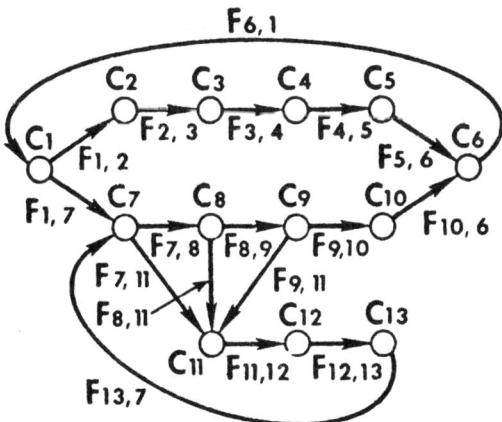

Figure 11-2. Conditional branching enables a robot to select one of several different state trajectories depending on sensed conditions.

More sophisticated yet are robots that can incorporate value of sensed variables into their data structure so as to track moving objects and operate upon or manipulate parts which are not precisely aligned to a known reference. When this is represented graphically, the result is that the states corresponding to program instructions become regions instead of points in state space. The extent of these regions corresponds to the range of the possible responses to the sensed variables.

As the length and complexity of branching in robot programs increases, it becomes tedious to write them as linear strings of instructions. Furthermore, it soon becomes evident to anyone actually writing such programs that the same substrings occur repeatedly. The obvious solution to this is to define macros, or subroutines, consisting of often used substrings. Of course, it is always possible to partition any particular task program into a set of macros as shown in Figure 11-3. If one can define a set of macros which can partition all programs within a certain class of tasks, it becomes possible to write programs for that class of tasks completely in terms of macros. The set of macros then becomes instruction set for a higher level programming language. Strings of macros represent robot programs written in the higher level language.

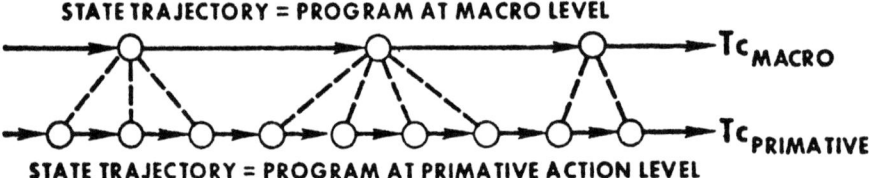

Figure 11-3. State trajectories corresponding to robot programs at two hierarchical levels.

Of course, strings of macros can themselves be partitioned into consistently recurring groups to form second level macros, and recurring groups of second level macros can be defined as third level macros, etc. In principal this process can be repeated any number of times to create a hierarchy wherein each level breaks a higher level input command into a sequence of subcommands to the next lower level. In this way a high level task is decomposed through a succession of hierarchical levels, until at the lowest level a string of action primitives produce the forces and motions to accomplish the high level task.

A four level hierarchical decomposition for a task ASSEMBLE (AB) is illustrated in Figure 11-4. On the left of Figure 11-4 is a hierarchy of computing modules, each representing a library of macros at the corresponding level. Each H module receives two sets of inputs; one a list, or vector, of command variables C, and the other a list, or vector, of feedback variables F. Each macro produces a state trajectory, and the H modules themselves are state machines. Each H module samples its inputs C and F and its own internal state at some time t = k and computes an output P. The output P becomes input to other H modules at time t = k + 1. Part of the F vector to any H modules consists of variables indicating the state of lower level modules. The rest of the F vector consists of sensory information from the environment filtered through a sensory processing hierarchy to that level.

Whether or not such an approach is practical depends upon how many macro names are required at each level to cover a class of tasks. At present this is still an open research issue. Clearly the list of names is task dependent. There are, however, a limited number of task types and a limited number of task decompositions required at each level by any single task. For example, the number of different types of elemental moves required for robots (or even humans) to perform routine mechanical assembly is not large. Systematic time and motion studies done for human workers (11) reveal a surprisingly small number of elemental movements corresponding to first level macros required for mechanical assembly of small components. There are only a few different types of reach, grasp, lift, transport, position, insert, twist, push, pull, release, etc. A list of parameters with each macro can specify where to reach, when to grasp, how far to twist, how hard to push, etc.

INCORPORATING REAL-TIME SENSORY FEEDBACK

From one perspective, the type of hierarchical task decomposition described above is nothing more than good top-down structured program design applied to robot tasks. Each macro represents a

164 MACHINE VISION

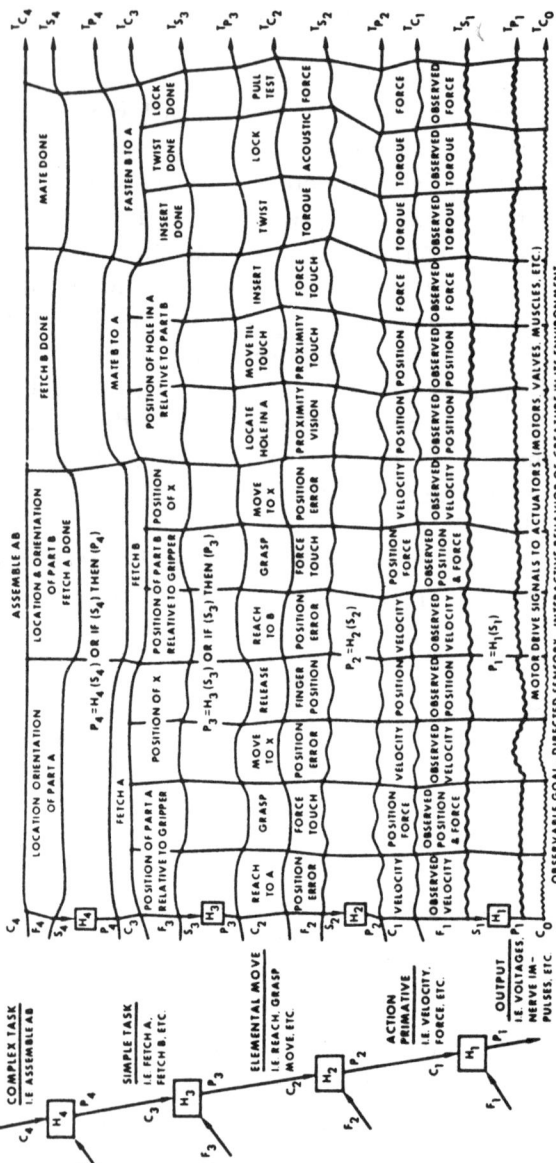

Figure 11-4. A hierarchy of state trajectories generated by a hierarchy of computing modules. Feedback at the various different levels produces sensory-interactive goal-directed behavior.

relatively short sequence of instructions and a limited and well structured set of branches. Programs at each level tend to be readable, understandable, and easy to debug and test for correctness.

The advantages of such a hierarchical decomposition extend far beyond programming convenience, however. The sophisticated realtime use of sensory measurement information for coping with uncertainty and recovering from errors requires that sensory data be able to interact with the control system at many different levels with many different constraints on speed and timing. For example, joint position, velocity, and sometimes force measurements are required at the lowest level in the hierarchy for servo feedback. These data require very little processing, but must be supplied without time delays of more than a few milliseconds. Visual depth (proximity) and information related to edges and surfaces are needed at the primitive action level of the hierarchy to compute offsets for gripping points. These data require a modest amount of processing and must be supplied within a few tenths of a second. Recognition of part position and orientation requires more processing and is needed at the elemental move level where time constraints are on the order of seconds. Recognition of parts and/or relationships between parts which may take several seconds is required for conditional branching at the simple task level.

Attempting to deal with this full range of sensory feedback in all of its possible combinations at a single level leads to extremely complex and inefficient programs. Sophisticated analysis of measurement data, particularly vision data (6, 7, 8, 9), is inherently a hierarchical process. Only if the control system is also partitioned into a hierarchy can the various levels of feedback information be introduced to the appropriate control levels in a simple and straight forward way.

MEASUREMENT IN AN ACTIVE PROCESS

In robotics the primary purpose of sensory feedback is to control action; but there are many different kinds of action. Different sensory information is required for different tasks and even different portions of the same task. As time varies and conditions change, different sensors, different resolutions, and different processing algorithms may be needed. The speed requirements of real-time control do not permit all resolutions and all processing algorithms to be applied all the time. Thus a real-time allocation of measurement and processing resources must be done.

Furthermore, measurement data often must be compared with expectations and hypotheses in order to detect missing objects or events, or to detect deviations from desired trajectories. Thus, the control system must have the capacity to tell the measurement system what to expect and when to expect it. This requires that there be links from the control hierarchy to the measurement hierarchy as well as the other way around. Figure 11-5 illustrates such a structure.

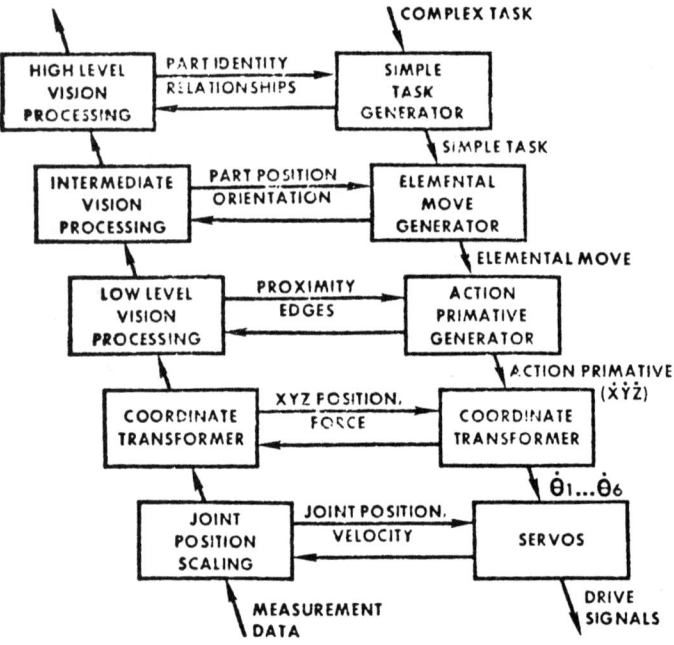

Figure 11-5. A cross-coupled measurement-control hierarchy. The links from left to right provide feedback for control. The links from right to left provide context and expectations for data processing. Here the lowest level of Figure 11-4 has been split into two levels: one for coordinate transformations between work space (or end effector space) and joint space, the other for servo computations on the joint positions and velocities.

These links may simply convey to the sensory processing modules what action is being taken so that the appropriate processing algorithms can be applied to the incoming data. However, in more sophisticated robot systems these links may also include data retrieved from associative memory or generated by mathematical models of the external environment. The entire set of links (including associative memory and mathematical models) represent a world model. Thus, the world model is itself a multilevel hierarchical entity.

As the task execution proceeds, the links from the control hierarchy comprising the world model produce sequences of expected data to the sensory processing modules at each level. Under ideal conditions these expected values will accurately predict the observed data. In other situations the processing modules at each level will detect deviations between the predictions and observations and produce error signals. These are the F vectors which modify the task decomposition at each level so as to maintain the robot performance within the limits of a success envelope.

In general, any deviation from the ideal task performance will be detected at the lowest level first. If the lowest level H function is capable of coping with that error, then no higher level action need be taken. However, if the error persists or grows in spite of the action of the lowest level H function, then it will be detected at the next higher level where a different macro may be selected. By this means it becomes possible to implement very sophisticated multilevel error correction procedures in a relatively stright forward manner.

SYSTEM IMPLEMENTATION

In our laboratory at NBS, we are in the process of implementing the type of cross-coupled measurement/control hierarchy described above. We have chosen to implement the hierarchy in a network of microcomputers because we believe this to be the best way to achieve low cost and upward compatibility. However, such a logical architecture can be implemented on a minicomputer. In fact, the first version of the NBS hierarchical control system (4) was implemented on a PDP 11/45.

The present system is a network of microcomputers with the architecture shown in Figure 11-6. Time is sliced into 20 millisecond increments. At the beginning of each increment each logical module reads its set of input values from the appropriate locations in common memory. It then computes its set of output values which it writes back into the common memory before the 20 millisecond interval ends. Any of the logical modules which do not complete their computations before the end of the 20 millisecond interval write extrapolated estimates of their output accompanied by a flag indicating that the data is extrapolated. The process then repeats.

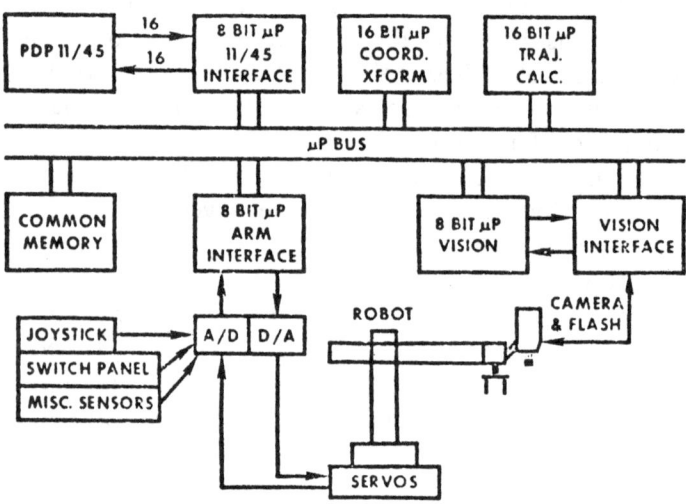

Figure 11-6. The NBS microcomputer network architecture for implementing a hierarchical robot control system.

Each logical module is thus a state machine whose output depends only on its present inputs and its present internal state. None of the logical modules admit any interrupts except the clock interrupt which signals the beginning and end of the 20 millisecond computation intervals. This simple module structure enormously simplifies the writing and debugging of software.

THE NBS VISION SYSTEM

The sensory side of the NBS hierarchical control system contains a vision system which uses active illumination to obtain depth information (10). A plane of light is generated by a photoflash tube and a cylindrical lens. This plane is projected into the field of view of a solid state 128 x 128 automation camera such that the distance to an illuminated surface can be directly computed by simple trigonometry. This camera and flash unit are fixed to the wrist of the robot manipulator.

The control hierarchy activates the vision system at specific points in a particular task execution. The control hierarchy also tells the vision software what type of object to expect and approximately how far away the object is expected to be. The vision software uses this information to select appropriate values for flash intensity and threshold and appropriate software algorithms for processing the visual data.

The vision processing modules either confirm the existence of the expected object and tell the control system where to move to approach it, or report that the expectation was incorrect.

At present, the NBS vision system interfaces with the control system primarily at the primitive action level for computing range and position of grip points and at the elemental move level for computing part orientation and approach paths. However, we are now in the process of adding new capabilities for part recognition at the simple task level.

COMMON MEMORY DATA TRANSFER

All communications of data from one module to another in the NBS hierarchical control system take place via a common memory "mail drop" system as shown in Figure 11-6. This system has a disadvantage in that it requires two data transfers to get information from one module to another. However, we believe this disadvantage is far overshadowed by the following advantages:

1. There are no communication protocols between computing modules, because modules do not talk directly to each other. Only one processor is allowed to write into any single loca-

tion in common memory. In each 20 millisecond time slice, all modules read from common memory before any are allowed to write their outputs back in.

2. The addition of each new state variable requires only a definition of where it is to be located in common memory so that the module which generates it knows where to write it, and the modules which read it know where to look. Thus, new microcomputers can easily be added, logical modules can be shifted from one microcomputer to another, new functions such as safety watchdogs, and even new sensors can be included with limited effect on the rest of the system. As long as the system bus has surplus capacity, the physical structure of the system can be reconfigured with few changes required in the software resident in the logical modules.

3. The common memory always contains a readily accessible map of the current state of the system. This makes it easy for a system monitor to trace the history of any or all of the state variables, to set break points, and to reason backwards to the source of program errors or faulty logic. This is extremely important in a sophisticated, real-time, sensory-interactive system in which many processes are going on in parallel at many different hierarchical levels.

The hierarchical decomposition and system implementation described in this chapter are still experimental and undoubtedly will change in many details as more experience is acquired. Nevertheless, results to date indicate that the modularity offered by this basic approach provides many advantages in software development, system extensibility, and ease of integration of sensory data.

References

[1] Albus, J.S., A.J. Barbera, J.M. Evans, and G.J. VanderBrug, "Control Concepts for Industrial Robots in an Automatic Factory," SME Technical Paper MS77-745, 1977

[2] Albus, J.S., "Mechanisms of Planning and Problem Solving in the Brain," Mathematical Biosciences 45:247-293 (1979)

[3] Barbera, A.J., J.S. Albus, and M.L. Fitzgerald, "Hierarchical Control of Robots Using Microcomputers," Proc. 9th Int. Symp. Indus. Robots, March 13-15, 1979, p. 405-422

[4] Barbera, A.J., "An Architecture for a Robot Hierarchical Control System," National Bureau of Standards Special Pub. 500-23, 1977

[5] Hasegawa, K., K. Suzuki, R. Masuda, and M. Kondo, "Programming and Teaching Method for Industrial Robot," Proc. 4th Int. Symp. Indus. Robots, Nov. 19-21, 1974, p. 301-310

[6] Ballard, D., C. Brown, and J. Feldman, "An Approach to Knowledge Directed Image Analysis," *Computer Vision Systems,* A. Hanson, E. Roseman (eds.) Academic Press, 1978, p. 271-281

[7] Hanson, A., and E. Riseman, "VISIONS: A Computer System for Interpreting Scenes," *Ibid* p. 303-333

[8] Reddy, R., "Pragmatic Aspects of Machine Vision," *Ibid* p. 89-98

[9] Tanimoto, S., "Regular Hierarchical Image and Processing Structures in Machine Vision," *Ibid* p. 165-174

[10] VanderBrug, G.J., J.S. Albus, and E. Barkmeyer, "A Vision System for Real Time Control of Robots," Proc. 9th Int. Symp. Indus. Robots, March 13-15, 1979, p. 213-232

[11] Yoosufani, A., and G. Boothroyd, "Design of Parts for Ease of Handling," National Science Foundation Grant #APR77 10197 Report #2 Chapter I, Review of Time and Motion Studies p. 1-101 (1978)

Appendix A
Machine Vision Companies

AEG Corporation
17177 North Laurel Park Drive
Livonia, MI 48152
(313) 597-2233

Arthur D. Little, Inc.
Acorn Park
Cambridge, MA 02140
(617) 864-5770

Adaptive Technologies
600 W. North Market Blvd., Suite 1
Sacramento, CA 95834
(916) 920-9119

Adept Technologies
1212 Bordeaux Drive
Sunnyvale, CA 94086
(408) 747-0111

Allen Bradley
1201 S. Second Street
Milwaukee, WI 53204
(414) 671-2000

American Industrial Vision Corp
174 Industry Drive
Pittsburg, PA 15275
(412) 787-8888

Anorad Corp.
110 Oser Ave.
Hauppauge, NY 11788
(516) 231-1990

Applied Intelligent Systems
110 Parkland Plaza
Ann Arbor, MI 48103
(313) 995-2035

Applied Scanning Technology
1988 Leghorn Street
Mountain View, CA 94043
(415) 967-4211

ASEA Robotics
16250 W. Glendale Dr.
New Berlin, WI 53151
(414) 785-3400

Associates & Ferran
P.O. Box 609
Wainscott, NY 11975
(516) 537-7800

Autoflex, Inc.
24655 Southfield Rd., Suite 110
Southfield, MI 48075
(800) 521-2144

Automatic Inspection
Devices, Inc.
One Seagate
Toledo, OH 43886
(419) 247-8128

Automatic Vision
1300 Richard Street
Vancouver, British Columbia, C,
V6B3G6

Automatix
1000 Tech Park Dr.
Billerica, MA 01821
(617) 667-7900

Gould
1200 West Colonial Dr.
Orlando, FL 32804
(305) 843-7030

Automation Systems, Inc.
340 Christian Rd.
Oxford, CT 06463
(203) 262-6166

Automation Engineering
11689 Sorrento Valley Rd.
San Diego, CA 92121
(619) 481-1295

Bahr Technologies
1842 Hoffman St.
Madison, WI 53704
(608) 244-0500

Beeco, Inc.
4175 Millersville Rd.
Indianapolis, IN 46205
(317) 597-1717

Brown & Sharpe/Validator
Systems
Presicion Park
North Kingston, RI 02852
(401) 886-7771

Cambridge Instruments
40 Robert Pitt Dr.
Monsey, NY 10952
(914) 356-3331

Cambridge Robotics
150 Coolidge Ave.
Watertown, MA 02172
(617) 926-7900

Cochlea
2284 Ringwood Ave., Bldg. C
San Jose, CA 95131
(408) 942-8228

Cognex
72 River Park St.
Needham, MA 02194
(617) 449-6030

Computer Systems
26401 Harper Ave.
St. Claire Shores, MI 48081
(313) 779-8700

Control Automation
P.O. Box 2304
Princeton, NJ 08540
(609) 799-6026

MANUFACTURERS OF MACHINE VISION SYSTEMS

Courser, Inc.
1262 Horseheads-Big Flats Road
Elmira, NY 14903
(607) 562-8413

CPI Technology (CENCIT)
12208 Mission Bottom Rd.
Hazlewood, MO 63042
(314) 213-1575

CR Technology
23062 LaLadena
Legruna Hills, CA 92653
(714) 859-4011

Cutler Hammer Div/Eaton Corp.
4201 N. 27th St.
Milwaukee, WI 53216
(414) 449-6000

Cybotech
P.O. Box 88514
Indianapolis, IN 46208
(317) 298-5890

Dac-Toyomenca America
9740 West Foster Ave.
Rosemount, IL 60018
(312) 992-2326

Data Sud Systems
2219 S. 48th St., Suite J
Tempe, AZ 85282
(602) 966-3953

DCI Corp.
110 S. Gold Dr.
Robbinsville, NJ 08691
(609) 587-9132

Diffracto
Box 2093
Windsor, Ontario Canada,
N8Y4R5
(519) 945-1591

Digital/Analog Design
530 Broadway
New York, NY 10012
(212) 966-0410

DIT-MC
5612 Brighton Ter.
Kansas City, MO 64130
(816) 444-9700

Durik/Kikusui
P.O. Box 580
Lakewood, NJ 08701
(201) 364-9700

Everett Charles
2887 N. Town Ave.
Pomona, CA 91767
(714) 621-9511

EGG/Reticon Div.
245 Potrero Ave.
Sunnyvale, CA 94086
(408) 738-4266

EISAI USA
20908 Higgins Ct.
Torrance, CA 90501
(213) 320-5790

Electra-Sol
2326 Fieldingwood Rd.
Maitland, FL 32751
(305) 339-0511

Electro-Optical Information
  Systems
710 Wilshire Blvd., Suite 501
Santa Monica, CA 90401
(213) 451-8566

ERIM
P.O. Box 8618
Ann Arbor, MI 48107

Excello/Raycon
77 Enterprise Dr.
Ann Arbor, MI 48103
(313) 769-2614

GEI
P.O. Box 10244
Winston-Salem, NC 27108
(919) 725-8494

General Electric
P.O. Box 17500
Orlando, FL 32860
(305) 889-1200

GMF Robotics
5200 New King St.
Troy, MI 48126
(313) 614-4140

General Numerics Corp.
390 Kent Ave.
Elk Grove Village, IL 60007
(312) 640-1595

Ham Industries
835 East Highland Road
Macedonia, OH 44056
(216) 467-4256

Hamamatsu Systems
332 Second Ave.
Waltham, MA 02154
(617) 890-3440

Hitachi, LTD
50 Prospect Ave.
Tarrytown, NY 10591
(914) 332-5800

Honeywell Visitronics
P.O. Box 5227
Denver, CO 80217
(303) 773-4437

Huges Aircraft Co./Industrial
  Products
6155 El Camino Real
Carlsbad, CA 92008
(714) 438-9191

Industrial Vision Systems
452 Chelmsford St.
Lowell, MA 01852
(617) 459-9000

MANUFACTURERS OF MACHINE VISION SYSTEMS

INEX Vision Systems
13327 US 19S
Clearwater, FL 33546
(813) 535-5502

INTEC
One Trefoil Dr., Trefoil Park
Trumbull, CT 06611
(203) 268-8000

Integrated Automation
1301 Harbour Bay Pkwy.
Alameda, CA 94501
(415) 769-5400

Intelledex
33840 Eastgate Circle
Corvallis, OR 97333
(503) 758-4700

Interactive Video Systems
358 Baker Ave.
Concord, MA 01742
(617) 371-0104

International Imaging Systems
1500 Buckeye Dr.
Milpitas, CA 95035
(408) 262-4444

International Robotmation/
   Intellig
2281 Las Palmas Dr.
Carlsbad, CA 92008
(619) 438-4424

ITEK
10 Maguire Rd.
Lexington, MA 02173
(617) 276-2000

ITMI
1000 Massachusetts Ave.
Cambridge, MA 02138
(617) 576-2585

ITRAN
670 N. Commercial St./Box 60
Manchester, NH 03105
(603) 669-6332

Jones & Lamson
501 Kings Mountain St.
York, SC 29745
(803) 684-9936

Key Image Data Systems
20100 Plummce St.
Chatsworth, CA 92311
(818) 993-1911

Key Technology, Inc.
P.O. Box 8
Milton-Freewater, OR 97862
(503) 938-5556

KLA Instruments Corp.
2051 Mission College Blvd.
Santa Clara, CA 95054
(408) 988-6100

Kulicke & Soffa (K&S)
104 Witmar Rd.
Horsham, PA 19044
(215) 443-7836

Lakso Company
P.O. Box 929
Leominister, MA 01453
(617) 537-8534

LNK Corp.
P.O. Box 136
College Park, MD 20740
(301) 927-3223

3M
8301 Greensboro Dr.
McClean, VA 22102
(703) 734-0300

Machine Intelligence
330 Potrero Ave.
Sunnyvale, CA 94086
(408) 737-7960

Machine Vision International
Burlington Center/325 East
    Eisenhower
Ann Arbor, MI 48104
(313) 996-8033

Mack WRP
3695 E. Industrial Dr.
Flagstaff, AR 86001
(602) 625-1120

Magnaflux
7300 W. Lawrence Ave.
Chicago, IL 60656
(312) 867-8000

Medar, Inc.
38700 Grand River Ave.
Farmington Hills, MI 48018

Micro-Poise
P.O. Box 88512
Indianapolis, IN 46208
(317) 298-5000

Micro-Vu
926 S. Lyon St., Box 15224
Santa Ana, CA 92704
(714) 547-6272

Mnemonics
P.O. Box C-131
Riverton, MJ 08077
(609) 786-8518

Monitor Automation
10180 Scripps Ranch Blvd.
San Diego, CA 92131
(419) 578-5060

NJSCORP (DAI NIPPON)
1633 Broadway, 15th Floor
New York, NY 10019
(212) 297-1880

ORS Automation
440 Wall St.
Princeton, NJ 08540
(609) 924-1667

Octek, Inc.
7 Corporate Place, S. Bedford St.
Burlington, MA 01803
(617) 273-0851

MANUFACTURERS OF MACHINE VISION SYSTEMS

OPCON
720 80th St., SW
Everett, WA 98203
(206) 353-0900

OPTEK
127 Holmes
Galena, OH
(614) 965-1819

Optical Gaging Products
850 Hudson Ave.
Rochester, NY 14621
(716) 544-0400

Optron Corp.
30 Hazel Terrace
Woodbridge, CT 06525
(203) 389-5384

Optrotech, Inc.
400 Amherst St.
Nashua, NH 03063
(603) 881-7041

Orbot Systems
98 South St.
Hopkinton, MA 01748
(617) 435-9355

Orthodyne Electronics
851 W. 185th St.
Costa Mesa, CA 92627
(714) 631-7800

Pattern Processing Technologies
511 11th Ave. South
Minneapolis, MN 55415
(612) 339-8488

Penn Video
929 Sweitzer Ave.
Akron, OH 44311
(216) 762-4840

Perceptron
23855 Research Dr.
Farmington Hills, MI 48024
(313) 478-7710

Photonic Automation
3633 W. MacArthur Blvd.
Suite Y-12
Santa Anna, CA 92704
(714) 546-6651

Photo Research/Kollmorgen
3099 N. Lima St.
Burbank, CA 91504
(818) 954-0104

Princeton Scientific Instruments
P.O. Box 252
Kingston, NJ 08528
(609) 921-6629

Production Automation Systems
Box 38, 511 11th Ave. S.
Minneapolis, MN 55415
(612) 338-4202

Prothon (Div. of Videotek)
199 Pomeroy Rd.
Parsippany, NJ 08054
(201) 887-8211

Rank Videometrix
9421 Winnetka Ave., Bldg. F
Chatsworth, CA 91311
(213) 701-6972

Raycon Corp.
77 Enterprise Ave.
Ann Arbor, MI 48103
(313) 769-2614

Robot Vision Systems
425 Rabro Drive East
Hauppauge, NY 11788
(516) 694-8910

Selcom/ Selective Electronics
Box 250
Valdese, NC 28690
(704) 874-2289

Sick Optik Electronik
2059 White Bear Ave.
St. Paul, MN 55109
(612) 777-9453

Simco-Remic
P.O. Box 1666
Medford, OR 97501
(503) 776-9800

Spectron Engineering
800 W. 9th Ave.
Denver, CO 80204
(303) 623-8987

Synthetic Vision Systems
2311 Green Rd.
Ann Arbor, MI 48105
(313) 665-1850

Technical Arts
180 Nickerson, Suite 102
Seattle, WA 98109
(206) 282-1703

Tencor Instruments
2426 Charleston Rd.
Mountain View, CA 94043
(415) 969-6767

Testerion/Mania
9645 Arrow Hwy.
Cucamonga, CA 91730
(714) 987-0025

TFI/QVAL Group
P.O. Box 1611
New Haven, CT 06506
(203) 934-5211

Unimation/Westinghouse
Shelter Rock Lane
Danbury, DT 06810
(203) 744-1800

Vanzetti Systems
111 Island St.
Stoughton, MA 02072
(617) 828-4650

Vektronics
5750 El Camino Real
Carlsbad, CA 92008
(619) 438-0992

Videk
343 State Street
Rochester, NY 14650

MANUFACTURERS OF MACHINE VISION SYSTEMS

View Engineering
1650 N. Voyager Ave.
Simi Valley, CA 93063
(805) 522-8439

Vuebotics
6086 Corte Del Cedro
Carlsbad, CA 92008
(619) 438-7994

Vision Systems Technologies
1532 S. Washington Ave.
Piscataway, NJ 08854
(201) 752-6700

Weldotron
1532 S. Washington Ave.
Piscataway, NJ 08854
(201) 752-6700

Appendix B
Glossary

aberration: failure of an optical lens to produce exact point-to-point correspondence between an object and its image.

accuracy: the degree to which the aritimetic average of a group of measurements conforms to the actual value or dimension.

aperture: of lens, opening that will pass light.

aspect ratio: the ratio of width to height for the frame of the television picture. The U.S. standard is 4:3.

backlighting: lighting an object so as to project a silhouette of the object.

bandwidth: the number of cycles per second expressing the difference between the lower and upper limiting frequencies of a frequency band.

barrel: distortion observed causing decreasing magnification from the center.

bimodal: a distribution of values with two peaks.

binary images: a black and white image obtained by thresholding a grey scale image reducing the image to black and white pixels.

bit slice: rudimentary building block type processor where one defines his own instruction set.

bit: binary digit—in image processing relates to image brightness—number of quantized levels.

blanking: the process whereby the beam in an image pickup tube is cut off during the retrace period.

blob: a group of connected pixels in a binary image.

blooming: the defocusing of regions of the picture where the brightness is at an excessive level.

boundaries: edge of a blob or object: See also edge.

boundary tracking (tracing): process which infers connectivity of pixels by following the edges of various blobs to determine their complete outlines.

burned-in image: (also called burn) an image which persists in a fixed position in the output signal of a camera tube after the camera has been turned to a different scene.

calibration: reconciliation to standard of measurement.

CCTV: abbreviation for Closed Circuit Television. A television system that does not broadcast TV signals but transmits them over a closed circuit.

centroid: midpoint of X and Y axis of an object; a blob feature measurement.

charge coupled device (CCD): a self-scanning semiconductor array that utilizes MOS technology, surface storage and information transfer by digital shift register techniques.

charge injection device (CID): a conductor-insulator-semiconductor structure that employs intracell charge transfer and charge injection to achieve an image sensing function.

chromaticity: see saturation

clustering algorithm: technique of processing image data based on evaluations of neighboring pixels, e.g., boundary tracking and connectivity analysis.

code reading: actual recognition of alphanumerics or other set of symbols, e.g., bar codes, UPC codes.

code verification: validation of alphanumeric data to assure conformance to qualitative standard—subjective.

GLOSSARY

color: process which stems from the selective absorption of certain wavelengths by an object.

computervision: see machine vision

connectivity (connectivity analysis): the process of building up individual blob descriptions on a pixel by pixel basis.

continuous motion: reflects condition where object cannot be stopped during the inspection process.

contrast: the range of difference between light and dark values in a picture, usually expressed as contrast ratio (the ratio between the maximum and minimum brightness value).

correlation: process of assessing the relationship between two or more models of an image.

dark current: the current that flows in a photoconductor when it is placed in total darkness.

depth of field: the in-focus range of a lens or optical system. It is measured from the distance behind an object to the distance in front of the object when the viewing lens shows the object to be in focus.

diffuse reflection: characteristic of light which leads to redirection over a range of angles from a surface on which it is incident.

diffuse transmission: characteristic of light that penetrates an object, scatters and emerges diffusely on the other side.

digitizing (digitization): process of converting an analog video image into digital brightness values that are assigned to each pixel in the digitized image.

edge: parts of an image characterized by rapid changes in intensity value.

edge detection: process of establishing edges in a scene by employing local operators that respond to the first or second derivative of the gray scale intensity in the neighborhood of each pixel.

An edge is detected when these derivatives exceed a given magnitude.

feature extraction: process of generating a set of descriptors for an image.

features: specific data elements describing the image content of a scene, such as edge point locations, centroid, etc.

field: one of the two equal parts into which a television frame is divided in an interlaced system of scanning.

field frequency: the number of fields transmitted per second in a television system. The U.S. standard is 60 fields per second. Also called field-repetition rate.

field of view: the maximum angle of view that can be seen through a lens or optical instrument.

filter: a transparent material characterized by selective absorption of light according to wavelength.

flaw detection: process of examining an object for unwanted features of an unknown shape at an unknown position.

focal length: of a lens, the distance from the focal point to the principal point of the lens.

focal plane: plane (through the focal point) at right angles to the principal axis of the lens.

focal point: the point at which a lens or mirror will focus parallel incident radiation.

frame: the total area, occupied by the television picture, which is scanned while the picture signal is not blanked.

frame frequency: the number of times per second that the frame is scanned. The U.S. standard is 30 frames per second.

frame-grabber: a device that stores entire picture frame, typically as a grey scale representation.

GLOSSARY 187

frame interlace: the method by which color and black and white sideband signals are interwoven within the same channel bandwidth.

frontlighting: projecting light on a surface of an object.

f/stop: speed or ability of a lens to pass light. FL/D.

gaging: in machine vision, non-contact dimensional examination of an object.

gamma: a numerical value, of the degree of contrast in a television picture, which is the exponent of that power law which is used to approximate the curve of output magnitude versus input magnitude over the region of interest.

geometric features: technique of classifying objects in a scene based on quantifiable features; e.g., centroid, moments, etc.

grey level: see bit

grey scale: description of contents of an image derived by conversion of analog video data from a sensor into proportional digital numbers.

grey scale vision: analysis of an image based on shade of grey content.

guidance: deriving properties in an image to describe position.

hierarchial approach: an image understanding system based on a series of ordered processing levels either "bottom-up" or "top-down"—hypothesize and test (train-by-showing or teaching).

histogram: a graph of a frequency distribution. In image analysis reflects frequency distribution of grey levels.

hue: the dominant wavelength of light representing the color of an object.

identification: determination of the identity of an object by reading symbols on the object.

image: a representation of physical object, an array of brightness values.

image analysis: process of generating a set of descriptors or features on which a decision about objects in an image is based.

image interpretation: process of building a description of a scene and then matching it against symbolic prototypes stored in a computer memory.

image preprocessing: reduction of image data to a manageable form.

image processing: operating on visual data to enhance content to improve visual decisions by an operator.

indexed motion: reflects condition where object can be stopped during the inspection process.

inspection: non-destructive examination of an object, with or without tools, to verify conformance to design.

interlaced scanning: a scanning process in which the distance from center to center of sucessively scanned lines is two or more times the normal line width, and in which the adjacent lines belong to different fields.

lag: in a television pickup tube, the persistence of the electrical charge image for two or more frames after excitation is removed.

laser scanner: technique of capturing image data by sweeping a laser over a scene and using a single detector to capture the variation in reflected light.

light: visually evaluated radiant energy of wavelengths from 380-770 nanometers.

light sectioning: see structured light

light striping: see structured light.

location: process of determining quantitatively the position of objects within a scene.

log spiral: non-linear parallel transform resulting in logarithmic polar mapping of a scene.

lookup table (LUT): memory that sets the input and output values for grey scale thresholding, windowing, inversion, and other display or analysis functions. Input values are for the storage of video unput pixels into image memory, and output values are for the display of stored pixels on the monitor. Also called translation table.

luminous intensity: brightness of a color.

machine vision: process of producing useful symbolic descriptions of a visual environment from image data—automatic interpretation of imagery to control a manufacturing process.

matched filtering: see template matching. Normally performed at the pixel level by cross correlation of an object template with an observed image field.

measurement: checking that parts are the right size—conform to specified dimensional tolerances.

median filtering: a method of local smoothing to replace each pixel with the median of the grey levels of neighborhood pixels.

model based analysis: evaluation of an image of an object based upon well defined geometric descriptions.

modulation: the process, or results of the process, whereby some characteristic of one signal is varied in accordance with another signal. The modulated signal is called the carrier. The carrier may be modulated in three fundamental ways; by varying the amplitude, called amplitude modulation; by varying the frequency, called frequency modulation; by varying the phase, called phase modulation.

moire: in television, a specialized optical device that makes it possible to use a single television camera in conjunction with one or more motion picture projectors and/or slide projectors in a film chain. The camera and projectors are in a fixed relationship, and prisms or special (dichroic) mirrors are used to provide smooth and instantaneous non-mechanical transition from one program source to the other.

negative image: a picture signal having a polarity which is opposite to normal polarity and which results in a picture in which the white and black areas are reversed.

neutral density filter: a light filter that reduces the intensity of light without changing the spectral distribution of the light.

noise removal: processing of image to remove unwanted artifacts from the image.

non-composite video: a video signal containing all information except sync.

NTSC: abbreviation for National Television Systems Committee. A committee that worked with the FCC in formulating standards for the present-day United States color television system.

optical processing: employs coherent optical interference methods to produce spatial Fourier transforms of objects then uses transforms to recognize objects.

optical scanner: device used to convert a picture into an array of numbers representing the positional distribution of optical density within the picture.

PAL: abbreviation for Phase Alternating Line System. A color television system in which the subcarrier derived from the color burst is inverted in phase from one line to the next in order to minimize errors in hue that may occur in color transmission.

pattern recognition: a technique that classifies images or objects within an image into predetermined categories, usually using statistical methods.

photoconductivity: photoconductivity deals with changes in the electrical conductivity of a material as a result of absorption of photons.

GLOSSARY

photoconductor: a device whose electrical resistance varies in relationship with exposure to light.

pincushion: distortion observed causing magnification increases as the distance from the center increases.

pipeline: refers to operating on an image in serial stages.

pixel: a spatial resolution element; smallest distinguishable and resolvable area in an image.

positional repeatability: specification that reflects how repeatably in space an object can be expected to pass (or be held in place) on inspection station; generally expressed in inches.

precision: the amount of spread of the distribution of measurements around some average value.

process control: operation which results in feedback to result in corrective action.

quality assurance: function of removing defective product from a manufacturing process.

recognition: determining an object's identity by using features of the object.

rectilinear transmission: characteristic of light observed when light passes through an object without diffusion.

refraction: bending of light as light travels at different speeds in one medium as compared to another.

repeatability: the ability to reproduce a result.

resolution: limit of detail that can be detected by a sensor.

robot guidance: process of image data to result in positional data fed back to robot to control end-effector position.

RS-170: Electronic Industries Association (EIA) standard governing monochrome television studio electrical signals. Specifies maximum amplitude of 1.4v, peak to peak, including synchronization pulses. Broadcast standard.

RS-232: EIA standard reflecting properties of serial communication link. RS-330: EIA standard governing closed-circuit television electrical signals. Specifies maximum amplitude of 1.0v, peak to peak, including synchronization pulses.

run length encoding: a data compression technique in which an image is raster scanned and only the lengths of "runs" of consecutive pixels with the same color are stored.

saturation: the pureness of a color.

scanning digitizer: see optical scanner.

seam track: non contact sensing method of providing feedback on the width of a gap to control a process, typically, arc welding.
- a) "thru the arc" — real time feedback in advance of the welding process—reflects influence of welding process itself on the gap.
- b) "a priori" — gap measurement before welding system generates gap data and path to follow by robot.

segmentation: the process of separating objects of interest from the rest of the scene or background—partitioning an image into various clusters.

shading: fidelity of grey level quantizing over the area of the image.

signal-to-noise ratio: the ratio of the peak value of the video signal to value of noise (usually expressed in decibels).

solid state camera: see CCD and CID

sort: determination which of a number of unknown objects or patterns is present—analogous to identification.

sorting: process of separating objects based on visual differences.

spatial noise: unwanted image artifacts in the image.

specular reflection: characteristic of light which leads to highly directional redirection, from a surface on which it is incident.

GLOSSARY

square-wave response: in image pickup tubes, the ratio of the peak-to-peak signal amplitude given by a test pattern consisting of alternate black and white bars of equal widths to the difference in signal between large-area blacks and large-area whites having the same illuminations as the black and white bars in the test pattern.

structured lighting: a special form of lighting which by triangulation makes it possible to measure distance from an object as well as infer shape and pose.

syntactic: a relationship that can be described by a set of rules, e.g., geometric remodelling.

template matching: process of comparing two images or image regions.

threshold: a specific pixel gray level value.

thresholding: segmentation process involving separation of a scene into regions of similar grey scale intensity.

throughput rate: generally refers to the number of objects to be examined per unit of time.

verification: an activity providing qualitative assurance that a fabrication or assembly process was successfully completed.

vidicon: an image pick up tube in which a charge density pattern is formed by photoconduction and stored on that surface of the photoconductor which is scanned by an electron beam, usually of low velocity electrons.

VLSI: very large grey scale integrated circuits.

windowing: technique of restricting examination of a scene to a subset of the scene.

Appendix C
Bibliography

Abbott, Edward H.
  Ford Motor Company
  Hegyi, Michelle A., and McCubbrey, David L.
  Environmental Research Institute of Michigan
  "Computer Algorithms for Visually Inspecting Thick Film Circuits"

Abraham, Richard G.; Yaroshuk, N.; and Beres, James F.
  Westinghouse Electric Corp.
  "Requirements Analysis and Justification of Intelligent Robots"

Abraham, R.G. and Shum, L.Y.
  Westinghouse Electric Corp., R.& D Center
  "Robot-Arc Welder with Contouring Teach Mode"

Agin, Gerald J.
  Stanford Research Institute
  "An Experimental Vision System for Industrial Application"

Albus, J.S. and Evans, J.M., Jr.
  U.S. National Bureau of Standards
  "A Hierarchical Structure for Robot Control"

Albus, Dr. James S.; Vanderbrug, G.J.; and Barkmeyer, E.
  U.S. National Bureau of Standards
  "A Vision System for Real Time Control of Robots"

Albus, Dr. James; Barbera, Dr. Anthony J.; and Vanderbrug, Dr.
  U.S. Dept. of Commerce National Bureau of Standards
  "Control Concepts for Industrial Robots in an Automatic Factory"

Armbruster, K.; Martini, P.; and Nehr, G.
  Universitat Karlsruhe (Germany)
  "A Very Fast Vision System for Recognizing Parts and Their Location and Orientation"

Bahn, Michael M. and Harned, John
  Autoflex, Inc.
  "Flexible Dimensional Gaging Systems"

Banks, David
  Object Recognition Systems, Inc.
  "Machine Vision—The Link Between Fixed and Flexible Automation"

Bardelli, R.; Dario, P.; DeRossi, D.; and Pinotti, P.C.
Centro "E Piaggio" (Pisa, Italy)
"Piezo- and Pyroelectric Polymers Skin-Like Tactile Sensors for Robots and Prostheses"

Barash, Dr. Moshe M.
Purdue University
"Production Technology Abroad—February, 1980"

Barash, Dr. Moshe M.
Purdue University
"Production Technology Abroad—August 1981"

Barbera, Dr. Anthony J.; Albus, Dr. James S.; and Evans, Dr. John M., Jr.
U.S. National Bureau of Standards
"Government Programs Relevant to the Advancement of Computer-Aided Manufacturing"

Barbera, Dr. Anthony J.; Albus, Dr. James S.; and Fitzgerald, M.L.
U.S. Dept. of Commerce National Bureau of Standards
"Hierarchical Control of Robots Using Microcomputers"

Benjamin, Harry L.
Centro Corporation
"The Development of a Production Robot Tactile Position Sensor"

Birla, Sushil K.
General Motors Technical Center
"Sensors for Adaptive Control and Machine Diagnostics"

Boardway, Robert A., and Frick, William N.
Ford Electronics and Refrigeration Corporation
"A Computer Vision System for Testing Electronic Clocks"

Botsco, Ronald J.
NDT Instruments, Inc.
"Ultrasonic Testing of Composites With High Resolution and Impedance Plane Techniques"

Brecker, J.N., and Shum, L.Y.
Westinghouse Electric Corp.
"Reducing Tool Wear With Air Gap Sensing"

Brennan, Michael A.
General Electric Co.
"Machine Vision and Robotic Testing"

Briggs, M. Darrell
Digital Equipment Corp.
"Printed Wiring Board Inspection"

Bowerman, E.R., and Kearns, R.F.
GTE Laboratories, Inc.
"Video Target Locator"

BIBLIOGRAPHY

Branaman, L.A.
    General Electric Company
    "Recent Applications of Electronic Vision to Non-Contact Inspection"

Briot, M.
    Laboratoire D'Automatique (France)
    "The Utilization of an "Artificial Skin" Sensor for the Identification of Solid Objects"

Brussel, H. Van, and Simons J.
    Katholieke Universiteit Leuven (Belgium)
    "The Adaptable Compliance Concept and Its Use for Automatic Assembly By Active Force Feedback Accommodations."

Budell, Robert A.
    Pontiac Motor Division, GM
    "Robotic Mig Welding Using Vision

Carlson, Mark J.
    Motorola, Inc.
    "Machine Vision: A Chip Placement Process Monitor"

Carter, Charles F., Jr.
    Cincinnati Milacron, Inc.
    "Machine Tool Characteristics Required for Versatility and Systems Use"

Casler, Richard J.
    Unimation, Inc.
    "Vision-Guided Robot Park Acquisition for Assembly and Packaging Applications"

Cassinis, Riccardo
    Istituto di Elettrotecnica Ed Elettronica Del
    "Sensing System in Supersigma Robot"

Chiesorin, P., and Lonardo, P.M.
    University of Genova
    "A New Sensor of Surface Roughness for Process Control System"

Chin, Roland; Harlow, Charles A.; and Dwyer, Samuel J., III
    University of Missouri-Columbia
    "Automation Inspection Techniques"

Chin, Ronald T.; Harlow, Charles A.; and Dwyer, Samuel J., III
    University of Missouri-Columbia
    "Automatic Visual Inspection of Printed Circuit Boards"

Clark, Lloyd; Webber, Neil; and Sutton, Terry
    International Harvester
    "Adaptive Control With the T3 Robot"

Clark, William J., and Frick, William N.
    Ford Electronics and Refrigeration Corporation
    "Robot Loading of P.C. Board Test Systems"

Colleen, Hans
 ASEA Incorporated
 "Adaptive Control: Giving Robots the Power to Cope"

Cook, George E.
 Venderbilt University and CRC Welding Systems, Inc.
 "Position Sensing With an Electric Arc"

Cooke, Dr. R.A.
 Trent Polytechnic, United Kingdom
 "Microcomputer Control of Free Ranging Robots"

Craig, J., and Cunningham, R.
 California Institute of Technology
 "Hand-Eye Control in a Robotic Assembly Task"

Cri, Hilmer
 Automatix Incorporated
 "Automatix Incorporated in Aerospace Application"

Cullen, Donald L.
 Autech Corporation
 "Miscellaneous Laser Measurements Including Strain and Product Dimension"

Cullen, Donald L.
 Autech Corporation
 "Precision Gauging Systems Utilizing Low Power Lasers"

Dario, P.; Domenici, C.; Bardelli, R.; De Rossi, D.; and Pinotti, P.C.
 Centro "E. Piaggio"
 "Piezoelectric Polymers: New Sensor Materials for Robotic Applications"

Del Gaudio, Italo, and D'Auria, Antonio
 Olivetti-O.S.A.I. (Italy)
 "Mechanical Behaviour of Sigma Robot"

Dessimoz, J.D.; Kunt, M.; and Zurcher, J.M.
 Swiss Federal Institute of Technology (Switzerland)
 "Recognition and Handling of Overlapping Industrial Parts"

Dixon, J.K.; Salazar, S.; and Slagle, J.R.
 Naval Research Lab
 "Research on Tactile Sensors for an Intelligent Naval Robot"

Falkman, Gerald A., and Murray, Lawrence A.
 Vuebotics Corporation
 "Robots and Vision: Where To?"

Fitzpatrick, David
 Object Recognition Systems, Inc.
 and
 Atkinson, Russ
 Affiliated Manufacturers, Inc.
 "AVISAS" Automatic Vision, Screen Alignment System

BIBLIOGRAPHY 199

Gaillet, Alain, and Reboulet, Claude
CERT
"A Isostatic Six-Component Force and Torque Sensor"

Geschke, Clifford C.
General Motors Corporation
"A Robot Task Using Visual Tracking"

Giralt, G.; Ghallab, M.; and Stuck, F.
Laboratoire D'Automatique (France)
"Object Identification and Sorting with an Optimal Sequential Pattern Recognition Method"

Gleason, Gerald J., and Agin, Gerald J.
SRI International
"A Modular Vision System for Sensor-Controlled Manipulation and Inspection"

Hahn, Robert S.
Hann Associates
"Precision Grinding Bodies of Revolution An Alternative to Diamond Turning"

Hall, Stephen M.
Industrial Measurement and Controls
"Vision Measurement Technology Using Photogrammetry for Quality Control in the Aerospace Industry"

Harmon, Leon D.
Case Western Reserve University
"Touch-Sensing Technology: A Review"

Hartwig, Glenn C.
Manufacturing Engineering
"Computer Vision: Unblinking Eyes for Quality Control"

Hayward, Vincent, and Paul, Richard P.
Purdue University
"Robot Manipulator Control Under Unix"

Hill, John and Park, William T.
SRI International
"Real Time Control of a Robot with a Mobile Camera"

Hoehn, Joseph
Automatic Inspection Devices, Inc.
"Vision Systems Applications at Owens-Illinois"

Holland, Steven W.
General Motors Corp.
"An Approach to Programmable Computer Vision for Robotics"

Hollar, Donald L., Jr.
    Bendix Corp.
    "Weld Process Monitoring Systems for Miniature Welding"

Holzer, Dr. Alexander J.
    Commonwealth Scientific Industrial Research Organization
    "A Robotic Cell for Application Evaluation and Training"

Hudson, David L.
    Octek, Inc.
    "Food for Thought and Appearance"

Ishlinsky, A. Yu.; Chernousko, F.L.; and Gradetsky, V.G.
    Institute for Problems of Mechanics/USSR Academy of Sciences
    "Some Problems of Mechanics and Control for Pneumatic Industrial Robots"

Jazitz, John A.
    Gulf & Western Co./E.W. Bliss Div.
    "Pressworking: Toward Higher Speeds, Greater Efficiencies"

Jeswiet, J. and Rice, W.B.
    Queen's University
    "The Design of a Sensor for Measuring Normal Pressure and Friction Stress in the Roll Gap During Cold Rolling"

Karg, R. and Lanz, O.E.
    BBC Brown Boveri Research Center (Switzerland)
    "Experimental Results With A Versatile Optoelectronic Sensor In Industrial Applications"

Kato, H.; Morinaga, S.; and Kato, T.
    Nagoya Municipal Industrial Research Institute (Japan)
    "A New Integrated Robot-Eye for Colour Discrimination"

Keil, Robert E.
    Honeywell—Corporate Production Technology Laboratory
    "Survey of Off-The-Shelf Imaging Systems"

Kelley, R.; Birk, J.; and Duncan, D.
    University of Rhode Island
    "A Robot System Which Feeds Workpieces Directly From Bins Into Machines"

Kelley, Robert K.; Birk, John R.; and Badami, V.
    University of Rhode Island
    "Workpiece Transportation by Robots Using Vision"

Kirsch, Jerry
    Auto-Place, Inc.
    "Progression of Intelligence in Limited Sequence Robots"

Kirsch, Kerry F.
    Copperweld Robotics
    "Pick-And-Place Robots in Small Part Assembly Systems"

Koch, Irvin D.
 Chevrolet Flint Manufacturing
 "Machine Vision Applications in Pressed Metal Parts Manufacture"

Kohno, M.; Takahashi, M.; and Isobe, M.
 No Author Affiliation
 "An Assembly Robot System with Twin Arm and Vision"

Kremers, Jan; Blahnik, Carl; Brain, Alfred; Cain, Ronald; Peppers, Norman; De Curtins, Jeff, and Meseguer, Jose
 SRI International
 "Development of a Machine-Vision-Based Robotic Arc Welding System"

Lapidus, Stanley N.
 Itran Corp.
 "New Techniques for Industrial Vision Inspection"

Levi, P. and Weirich, E.
 University Karlsruhe
 "Differential Reflectance Functions and Their Use for Surface Identification"

Makhlin, Arkand G.
 Digital Equipment Corporation
 "Grey Scale Robot Vision for Real-Time Inspection and Assembly"

Malinen, Pekka and Miemi, Antti
 Helsinki University of Technology (Finland)
 "Reduction of Visual Data by a Program Controlled Interface for Computerized Manipulation"

Mallick, G.T., Jr.
 Westinghouse Electric Corp.
 "Acoustic Die Monitoring"

Mathias, Richard A.
 Camtech, Inc.
 "Shop Floor Data Monitoring and Processing Requirements for Generative Cutter Path Selection and Machining Rate Optimization"

Mathias, Richard A.
 Camtech Inc.
 "Unmanned Collection and Processing of Shop Floor Machining Data"

McCraley, Michael T.
 Control Automation, Inc.
 "Lead Frame Inspection—High Speed Analysis—Low-Cost Solution"

Mc'Ghie, Dennis and Hill, John W.
 SRI International
 "Vision Controlled Subassembly Station"

McKinney, Raymond H.
    Object Recognition Systems, Inc.
    "Development of a Specification for a Machine Vision System"

Mersch, Steven H.
    University of Dayton Research Institute
    "Polarized Lighting for Machine Vision Applications"

Meyer, John D.
    Tech. Tran. Corp.
    "The Role of Machine Vision in Flexible Manufacturing Systems"

Miller, John W.V.
    Owens-Illinois, Inc.
    "A Real-Time System for the Automatic Inspection of Uniform Field Objects"

Movich, Richard C.
    Lockheed California Company
    "Robotic Drilling and Riveting Using Computer Vision"

Nagel, Roger N.; Vanderbrug, Gordon J.; and Albus, Dr. James S.
    U.S. National Bureau of Standards
    "Experiments in Part Acquisition Using Robot Vision"

Nogami, M. and Watanabe, M.
    Mitsubishi Electric Corporation
    "Flexibility in Production Line"

Nowak, Glenn
    Copperweld Robotics
    "The Advent of Machine Vision Systems"

Olson, W. Richard
    Honeywell, Inc.
    Raymond, Chuck and Donath, Max
    University of Minnesota
    "Problems of Vision-Directed Robots in an Unstructured Parts Handling Environment"

Pekelharing, A.J. and Orelio, J.M.B.
    Delft University of Technology (The Netherlands)
    "When Does the Cutting Tool Crack?"

Porter, G.B., III
    General Electric Company
    "Automatic Inspection of Surface Features Using Functional Approximation of Grey Level Data"

Porter, Gilbert G. III
    General Electric
    "A Non-Contact Visual Profilimeter for Automatic Inspection"

Potter, Ronald D.
> Auto-Place Inc.
> "Applications of Industrial Robots with Visual Feedback"

Potter, Ronald D.
> Auto-Place, Inc.
> "Applications of Low Cost Robots"

Potter, Ronald D.
> Auto-Place, Inc.
> "Practical Applications of a Limited Sequence Robot"

Pryor, Tim and Pastorius, Walt
> Diffracto Limited
> "Applications of Machine Vision to Parts Inspection and Machine Control in the Piece Part Manufacturing Industries"

Pryor, Tim
> Diffracto Limited
> "Electro-Optical Inspection and Machine Vision"

Rioux, Marc
> Conseil National De Recherches
> "3-D Camera Based on Synchronized Scanning"

Robertson, Gordon I.
> Control Automation, Inc.
> "Heirarchical Control of Intelligent Robot and Vision Allows Plug-In System Integration"

Rosen, C.A. and Gleason, G.J.
> Machine Intelligence Corporation
> "Evaluation of Performance of Machine Vision Systems"

Rovetta, Alberto
> Politecnico of Milan
> "A New Robot with Voice, Hearing, Vision, Touch, Grasping, Controlled by One Microprocessor, with Mechanical and Electronic Integrated Design"

Roy, John
> Zygo Corp.
> "Laser Based In-Process Dimension Measurement and Control"

Sakai, I. and Kumazawa, T.
> Nagoya University
> Shingu, H.
> Aichi Institute of Technology
> "Approach and Plan: Most Suitable Control of Grasping in Industrial Robot"

Saladino, John
> General Elctric Company
> "Upset Forging with Industrial Robots"

Schraft, R.D.; Schweizer, M.; Abele, E.; and Sturz, W.
    Fraunhofer Inst. For Mfg. Engineering And Automation (IPA)
    Federal Republic of Germany
    "Application of Sensor-Controlled Robots for Fettling of Castings"

Schroeder, H.E.
    EG&G Reticon
    "Circular Scanning and Its Relationship to Machine Vision"

Selfridge, P.G.
    Bell Laboratories
    "An Interactive Software System for Developing Robot Vision Algorithms"

Seltzer, Donald S.
    Charles Stark Draper Laboratory, Inc.
    "Tactile Sensory Feedback for Difficult Robot Tasks"

Seltzer, Donald S.
    Charles Stark Draper Laboratory, Inc.
    "Use of Sensory Information for Improved Robot Learning"

Seres, David A.; Kelly, Robert K.; and Birk, John R.
    University of Rhode Island
    "Visual Robot Instruction"

Shillman, Robert J.
    Cognex Corporation
    "Automated Part Recognition Via Optical Character Recognition"

Sim, Jack
    White-Sundstrand Machine Tool, Inc.
    "In Process Computer Control"

Skoog, Hans
    ASEA AB (Sweden)
    "Adaptivity—As Applied to Industrial Robot"

Smith, Bradford M.
    U.S. National Bureau of Standards
    "New Dimensions in Automation Technology"

Smith, Bruce S. and Petersson, Christer U.
    ASEA Robotics
    "An Integrated Robot Vision System for Industrial Use"

Society of Manufacturing Engineers Editorial Staff
    Robotics Today
    "Automatix Develops Assembly System for DEC"

Society of Manufacturing Engineers Editorial Staff
    Robotics Today
    "Composites and Robotics: Pushing Technology Forward"

Society of Manufacturing Engineers Editorial Staff
  Manufacturing Engineering
  "Production-Oriented Safety Device for Pressworking"

Society of Manufacturing Engineers Editorial Staff
  Robotics Today
  "Promoting Automation at Chesebrough-Pond's"

Society of Manufacturing Engineers Staff
  "Vision Systems"

Society of Manufacturing Engineers
  Videotape
  "Robots VI: Tomorrow's Technology on Display"

Spur, G.; Auer, B.H.; and Weisser, W.
  Technical University of Berlin
  "Handling Automation for Two Milling Machine Tools"

Stauffer, Robert
  Editor
  "Vision and Sensing"

Stauffer, Robert N.
  Robotics Today
  "Robots VI Triggers Spurt in New Technology"

Stauffer, Robert N.
  Robotics Today
  "The 11th International Symposium on Industrial Robots"

Stockman, George C.
  Michigan State University
  "Circular Scanning for Part Recognition and Orientation

Stute, G. and Erne, H.
  Universitaet Stuttgart (West Germany)
  "The Control Design of an Industrial Robot With Advanced Tactile Sensitivity"

Svetkoff, Donald J.; Candlish, John B.; and Vanatta, Peter W.
  Environmental Research Institute of Michigan
  "High-Resolution Imaging for Automatic Inspection of Multi-Layer Thick Film Circuits"

Tanaka, Masato; Suzuki, Takeo; and Seryu, Mitsuo
  Yaskawa Electric Manufacturing Co., Ltd.
  "Flexible Handling Systems for Small Parts"

Tanner, William R.
  Editor
  "Inspection"

Trombly, John E.
    Octek, Inc.
    "A New Approach to Machine Vision Simplifies Application Development"

Trombly, John E.
    Octek, Inc.
    "Small Part Sorting & Inspection Using Machine Vision"

Tucker, B.J.
    EMR Photoelectric
    "Quality and Optical Scanning"

Tutching, John
    John Tutching & Company
    "The Technological, Market, and Applications Environment for Machine Vision Systems 1984-1987: An Analysis, Study, and Assessment of Technological Transition, Applications Development, Markets and Related Factors"

Ueda, M.; Matsuda, F.; and Matsuyama, S.
    Nagoya University (Japan)
    "A Simple Distance Sensor and a New Mini-Computer System"

Umetani, Yoji and Taguchi, Kan
    Tokyo Institute of Technology (Japan)
    "Feature Properties to Discriminate Complex Shapes"

Vanderbrug, Gordon and Wilt, Donald
    Automatix, Inc.
    Davis, Jim
    Digital Equipment Corporation
    "Robotic Assembly of Keycaps to Keyboard Arrays"

Vasilash, Gary S.
    Robotics Today
    "The Cellular Approach to Image Processing"

Villers, Philippe
    Automatix Inc.
    "Recent Proliferation of Industrial Artificial Vision Applications"

Voltz, R.A.; Mudge, T.N.; and Gal, D.A.
    Center for Robotics and Integrated Manufacturing
    "Using ADA as a Robot System Programming Language"

Vranish, John M.
    Naval Surface Weapons Center
    "Magnetoelastic Force Feedback Sensors for Robots and Machine Tools"

Vranish, John M.
    Naval Surface Weapons Center
    "The Robotic Deriveter—Systems Concept"

Waldman, Harry
    Manufacturing Engineering
    "Automatic Assembly: Making Use of Probes and Vision Systems"

Walker, David E.
    Argonne National Laboratory
    "Brazed Thermocouple Pass-Through for Sodium Service in a Liquid-Metal-Cooled Fast Breeder Reactor"

Walter, Wayne W.
    Rochester Institute of Technology
    "Applying Robotic Vision for Assembly and Fabrication"

Ward, M.R.; Rossol, L.; and Holland, S.W.
    General Motors Corporation
    "Consight: A Practical Vision-Based Robot Guidance System"

Warnecke, Dr. H.J.; Schraft, Dr. R.D.; and Brodeck, B.
    Fraunhofer-Institute for Production and Automation
    "Pilot Work Site With Industrial Robots"

Watson, Paul and Drake, Samuel H.
    The Charles Stark Draper Laboratory
    "Pedestal and Wrist Force Sensors for Automatic Assembly"

Wheatley, Thomas; Albus, James; and Nagel, Roger
    U.S. National Bureau of Standards
    "Proceedings of NBS/Air Force ICAM Workshop on Robot Interfaces"

Will, P.M.
    IBM Corporation, Thomas J. Watson Research Center
    "Computer Controlled Mechanical Assembly"

Wilson, Kenneth R.
    R.J. Wilson Associates, Inc.
    "Fiber Optics: Practical Vision for the Robot"

Wittels, Norman and Libby, Charles J.
    Automatix, Inc.
    "Vision Aided Arc Welding"

Yu, Lee-Chi
    Beijing Machine Tool Research Institue, China
    Zhang, Xiaqzy
    Shenyang Industrial College of Machine Tool Co., China
    "The Positioning Accuracy of Industrial Robots"

Index

A
Alphanumeric character recognition 78
Applications 61, 79
Assembly 112, 115, 134

B
Benefits 2, 51, 54
Binary Processing 26
Bin Picking 119

C
CAD/CAM 5
Cameras 9-14, 48
Castings 69, 70
Charge Coupled Devices 10, 17, 48
Color sensing 36, 49
Connectivity Analysis 26
Containers 61
Computer interface 29

E
Edge Detection 30

F
Food industry 66
Future 29

G
Gaging 62, 65
Glass inspection 69
Grading 66
Gray Scale 27, 110

H
Hardware 9, 108
Hierarchial 160
Histogram 6, 27

J
Justification 2, 51

L
Label Inspection 61
Languages 29, 132
Lasers 34
Lighting 17, 110

M
Model-based vision 36, 135, 137

P
Pixels 9, 15, 16, 48
Printed circuit board inspection 72, 114
Processing time 15, 49

Q
Quality assurance 5

R
Range imaging sensors 34
Request for proposal 57
Resolution 12, 48
Robots 89, 111, 132, 139, 145, 159
Run-length encoding 30

T
3-dimensional vision 62

V
Vendors 51, 173
Vidicon 9

W
Windowing 30